# A Guide to Graph Colouring

R.M.R. Lewis

# A Guide to Graph Colouring

## Algorithms and Applications

 Springer

R.M.R. Lewis
Cardiff School of Mathematics
Cardiff University
Cardiff
UK

ISBN 978-3-319-37282-2        ISBN 978-3-319-25730-3 (eBook)
DOI 10.1007/978-3-319-25730-3

Springer Cham Heidelberg New York Dordrecht London
© Springer International Publishing Switzerland 2016
Softcover re-print of the Hardcover 1st edition 2016

Printed on acid-free paper

Springer International Publishing AG Switzerland is part of Springer Science+Business Media
(www.springer.com)

*For Fifi, Maiwen, Aoibh, and Macsen.*
*Gyda cariad.*

# Preface

Graph colouring is one of those rare examples in the mathematical sciences of a problem that is very easy to state and visualise, but that has many aspects that are exceptionally difficult to solve. Indeed, it took more than 160 years and the collective efforts of some of the most brilliant minds in nineteenth and twentieth century mathematics just to prove the simple sounding proposition that "four colours are sufficient to properly colour the vertices of a planar graph".

Ever since the notion of "colouring" graphs was first introduced by Frances Guthrie in the mid-1800s, research into this problem area has focussed mostly on its many theoretical aspects, particularly concerning statements on the chromatic number for specific topologies such as planar graphs, line graphs, random graphs, critical graphs, triangle free graphs, and perfect graphs. Excellent reviews on these matters, together with a comprehensive list of open problems in the field of graph colouring, can be found in the books of Jensen and Toft (1994) and Beineke and Wilson (2015).

In this book, our aim is to examine graph colouring as an *algorithmic* problem, with a strong emphasis on practical applications. In particular, we take some time to describe and analyse some of the best-known algorithms for colouring arbitrary graphs and focus on issues such as (a) whether these algorithms are able to provide optimal solutions in some cases, (b) how they perform on graphs where the chromatic number is unknown, and (c) whether they are able to produce better solutions than other algorithms for certain types of graphs, and why.

This book also devotes a lot of effort into looking at many of the real-world operational research problems that can be tackled using graph colouring techniques. These include the seemingly disparate problem areas of producing sports schedules, solving Sudoku puzzles, checking for short circuits on printed circuit boards, assigning taxis to customer requests, timetabling lectures at a university, finding good seating plans for guests at a wedding, and assigning computer programming variables to computer registers.

This book is written with the presumption that the reader has no previous experience in graph colouring, or graph theory more generally. However, an elementary knowledge of the notation and concepts surrounding sets, matrices, and enumerative combinatorics (particularly combinations and permutations) is assumed. The initial sections of Chapter 1 are kept deliberately light, giving a brief tour of the graph colouring problem using minimal jargon and plenty of illustrated examples. Later sections of this chapter then go on to look at the problem from an algorithmic point of view, looking particularly at why this problem is considered "intractable" in the general case, helping to set the ground for the remaining chapters.

Chapter 2 of this book looks at three different well-established constructive algorithms for the graph colouring problem, namely the GREEDY, DSATUR, and RLF algorithms. The various features of these algorithms are analysed and their performance (in terms of running times and solution quality) is then compared across a large set of problem instances. A number of bounds on the chromatic number are also stated and proved.

Chapters 3 and 4 then go on to look at some of the best-known algorithms for the general graph colouring problem. Chapter 3 presents more of an overview of this area and highlights many of the techniques that can be used for the problem, including backtracking algorithms, integer programming methods, and metaheuristics. Ways in which problem sizes can be reduced are also considered. Chapter 4 then goes on to give an in-depth analysis of six such algorithms, describing their relevant features, and comparing their performance on a wide range of different graph types. Portions of this chapter are based on the research originally published by Lewis et al. (2012).

Chapter 5 considers a number of example problems, both theoretical and practical, that can be expressed using graph colouring principles. Initial sections focus on special cases of the graph colouring problem, including map colouring (together with a history of the Four Colour Theorem), edge colouring, and solving Latin squares and Sudoku puzzles. The problems of colouring graphs where only limited information about a graph is known, or where a graph is subject to change over time, are also considered, as are some natural extensions to graph colouring such as list colouring, equitable graph colouring and weighted graph colouring.

The final three chapters of this book look at three separate case studies in which graph colouring algorithms have been used to solve real-world practical problems, namely the design of seating plans for large gatherings, creating schedules for sports competitions (Lewis and Thompson, 2010), and timetabling events at educational establishments (Lewis and Thompson, 2015). These three chapters are written so that, to a large extent, they can be read independently of the other chapters of this book, though obviously a full understanding of their content will only follow by referring to the relevant sections as instructed by the text.

## A Note on Pseudocode and Notation

While many of the algorithms featured in this book are described within the main text, others are more conveniently defined using pseudocode. The benefit of pseudocode is that it enables readers to concentrate on the algorithmic process without worrying about the syntactic details of any particular programming language. Our pseudocode style is based on that of the seminal textbook *Introduction to Algorithms* by Cormen, Leiserson, Rivest and Stein, often simply known as the "The Big Book of Algorithms" (Cormen et al., 2009). This particular pseudocode style makes use of all the usual programming constructs such as while-loops, for-loops, if-else statements, break statements, and so on, with indentation being used to indicate their scope. To avoid confusion, different symbols are also used for assignment and equality operators. For assignment, a left arrow ($\leftarrow$) is used. So, for example, the statement $x \leftarrow 10$ should be read as "$x$ becomes equal to 10", or "let $x$ be equal to 10". On the other hand, an equals symbol is used only for equality *testing*; hence a statement such as $x = 10$ will only evaluate to true or false ($x$ is either equal to 10, or it is not).

All other notation used within this book is defined as and when the necessary concepts arise. Throughout the text, the notation $G = (V, E)$ is used to denote a graph $G$ comprising a "vertex set" $V$ and an "edge set" $E$. The number of vertices and edges in a graph are denoted by $n$ and $m$ respectively. The colour of a particular vertex $v \in V$ is written $c(v)$, while a candidate solution to a graph colouring problem is usually defined as a partition of the vertices into $k$ subsets $S = \{S_1, S_2, \ldots, S_k\}$. Further details can be found in the various definitions within Chapters 1 and 2.

## Additional Resources

This book is accompanied by a suite of nine graph colouring algorithms that can be downloaded from www.rhydlewis.eu/resources/gCol.zip. Each of these heuristic-based algorithms are analysed in detail in the text and are also compared and contrasted empirically through extensive experimentation. These implementations are written in C++ and have been successfully compiled on a number of different compilers and platforms. (See Appendix A.1 for further details.) Readers are invited to experiment with these algorithms as they make their way through this book. Any queries should be addressed to the author.

In addition to this suite, this book's appendix also contains information on how graph colouring problems might be solved using commercial linear programming software and also via the free mathematical software Sage. Finally, an online implementation of the table planning algorithm presented in Chapter 6 can also be accessed at www.weddingseatplanner.com.

Cardiff University, Wales.                                                               *Rhyd Lewis*
August 2015

# Contents

# Chapter 1
# Introduction to Graph Colouring

In mathematics, a graph can be thought of as a set of objects in which some pairs of objects are connected by links. The interconnected objects are usually called *vertices*, with the links connecting pairs of vertices termed *edges*. Graphs can be used to model a surprisingly large number of problem areas, including social networking, chemistry, scheduling, parcel delivery, satellite navigation, electrical engineering, and computer networking. In this chapter we introduce the graph colouring problem and give a number of examples of where it is encountered in real-world situations. Statements on the complexity of the problem are also made.

The graph colouring problem is one of the most famous problems in the field of graph theory and has a long and illustrious history. In a nutshell it asks, given any graph, how might we go about assigning "colours" to all of its vertices so that (a) no vertices joined by an edge are given the same colour, and (b) the number of different colours used is minimised?

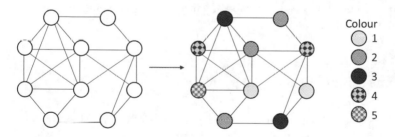

**Fig. 1.1** A small graph (a), and corresponding 5-colouring (b)

Figure 1.1 shows a picture of a graph with ten vertices (the circles), and 21 edges (the lines connecting the circles). It also shows an example colouring of this graph that uses five different colours. We can call this solution a "proper" colouring because all pairs of vertices joined by edges have been assigned to different colours, as required by the problem. Specifically, two vertices have been assigned to colour 1, three vertices to colour 2, two vertices to colour 3, two vertices to colour 4, and one vertex to colour 5.

© Springer International Publishing Switzerland 2016
R.M.R. Lewis, *A Guide to Graph Colouring*,
DOI 10.1007/978-3-319-25730-3_1

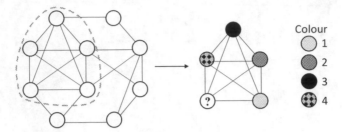

**Fig. 1.2** If we extract the vertices in the dotted circle, we are left with a subgraph that clearly needs more than four colours

Actually, this solution is not the only possible 5-colouring for this example graph. For example, swapping the colours of the bottom two vertices in the figure would give us a different proper 5-colouring. It is also possible to colour the graph with anything between six and ten colours (where ten is the number of vertices in the graph), because assigning a vertex to an additional, newly created, colour still ensures that the colouring remains proper.

But what if we wanted to colour this graph using *fewer* than five colours? Is this possible? To answer this question, consider Figure 1.2, where the dotted line indicates a selected portion of the graph. When we remove everything from outside this selection, we are left with a subgraph containing just five vertices. Importantly, we can see that every pair of vertices in this subgraph has an edge between them. If we were to have only four colours available to us, as indicated in the figure we would be unable to properly colour this subgraph, since its five vertices all need to be assigned to a different colour in this instance. This allows us to conclude that the solution in Figure 1.1 is actually *optimal*, since there is no solution available that uses fewer than five colours.

## 1.1 Some Simple Practical Applications

Let us now consider four simple practical applications of graph colouring to further illustrate the underlying concepts of the problem.

### 1.1.1 A Team Building Exercise

An instructive way to visualise the graph colouring problem is to imagine the vertices of a graph as a set of "items" that need to be divided into "groups". As an example, imagine we have a set of university students that we want to split into groups for a team building exercise. In addition, imagine we are interested in dividing the students so that no student is put in a group containing one or more of his friends, and so that the number of groups used is minimal. How might this be done?

Consider the example given in the table in Figure 1.3(a), where we have a list of eight students with names A through to H, together with information on who their friends are. From this information we can see that student A is friends with three students (B, C and G), student B is friends with four students (A, C, E, and F), and so on. Note that the information in this table is "symmetric" in that if student $x$ lists student $y$ as one of his friends, then student $y$ also does the same with student $x$. This sort of relationship occurs in social networks such as Facebook, where two people are only considered friends if both parties agree to be friends in advance. An illustration of this example in graph form is also given in the figure.

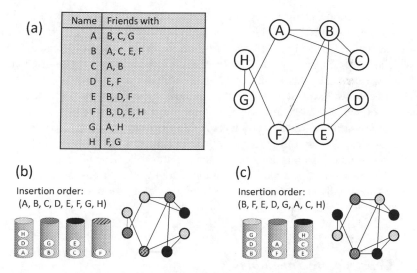

**Fig. 1.3** Illustration of how proper 5- and 4-colourings can be constructed from the same graph

Let us now attempt to split the eight students of this problem into groups so that each student is put into a different group to that of his friends'. A simple method to do this might be to take the students one by one in alphabetical order and assign them to the first group where none of their friends are currently placed. Walking through the process, we start by taking student A and assigning him to the first group. Next, we take student B and see that he is friends with someone in the first group (student A), and so we put him into the second group. Taking student C next, we notice that he is friends with someone in the first group (student A) and also the second group (student B), meaning that he must now be assigned to a third group. At this point we have only considered three students, yet we have created three separate groups. What about the next student? Looking at the information we can see that student D is only friends with E and F, allowing us to place him into the first group alongside student A. Following this, student E cannot be assigned to the first group because he is friends with D, but can be assigned to the second. Continuing this process for all eight students gives us the solution shown in Figure 1.3(b). This solution uses four groups, and also involves student F being assigned to a group by himself.

Can we do any better than this? By inspecting the graph in Figure 1.3(a), we can see that there are three separate cases where three students are all friends with one another. Specifically, these are students A, B, and C; students B, E, and F; and students D, E, and F. The edges between these triplets of students form triangles in the graph. Because of these mutual friendships, in each case these collections of three students will need to be assigned to different groups, implying that *at least* three groups will be needed in any valid solution. However, by visually inspecting the graph we can see that there is no occurrence of *four* students all being friends with one another. This hints that we may not necessarily need to use four groups in a solution.

In fact, a solution using three groups *is* actually possible in this case as Figure 1.3(c) demonstrates. This solution has been achieved using the same assignment process as before but using a different ordering of the students, as indicated. Since we have already deduced that at least three groups are required for this particular problem, we can conclude that this solution is *optimal*.

The process we have used to form the solutions shown Figures 1.3(b) and (c) is generally known as the GREEDY algorithm for graph colouring, and we have seen that the ordering of the vertices (students in this case) can influence the number of colours (groups) that are ultimately used in the solution it produces. The GREEDY algorithm and its extensions are a fundamental part of the field of graph colouring and will be considered further in later chapters. Among other things, we will demonstrate that there will always be at least one ordering of the vertices that, when used with the GREEDY algorithm, will result in an optimal solution.

### 1.1.2 Constructing Timetables

A second important application of graph colouring arises in the production of timetables at colleges and universities. In these problems we are given a set of "events", such as lectures, exams, classroom sessions, together with a set of "timeslots" (e.g., Monday 09:00–10:00, Monday 10:00–11:00 and so on). Our task is to then assign the events to the timeslots in accordance with a set of constraints. One of the most important of these constraints is what is often known as the "event-clash" constraint. This specifies that if a person (or some other resource of which there is only one) is required to be present in a pair of events, then these events must not be assigned to the same timeslot since such an assignment will result in this person/resource having to be in two places at once.

Timetabling problems can be easily converted into an equivalent graph colouring problem by considering each event as a vertex, and then adding edges between any vertex pairs that are subject to an event clash constraint. Each timeslot available in the timetable then corresponds to a colour, and the task is to find a colouring such that the number of colours is no larger than the number of available timeslots.

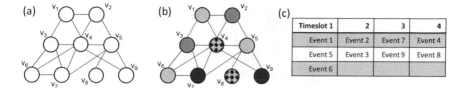

**Fig. 1.4** A small timetabling problem (a), a feasible 4-colouring (b), and its corresponding timetable solution using four timeslots (c)

Figure 1.4 shows an example timetabling problem expressed as a graph colouring problem. Here we have nine events which we have managed to timetable into four timeslots. In this case, three events have been scheduled into timeslot 1, and two events have been scheduled into each of the remaining three. In practice, assuming that only one event can take place in a room at any one time, we would also need to ensure that three rooms are available during timeslot 1. If only two rooms are available in each timeslot, then an extra timeslot might need to be added to the timetable.

It should be noted that timetabling problems can often vary a great deal between educational institutions, and can also be subject to a wide range of additional constraints beyond the event-clash constraint mentioned above. Many of these will be examined further in Chapter 8.

### 1.1.3 Scheduling Taxis

A third example of how graph colouring can be used to solve real-world problems arises in the scheduling of tasks that each have a start and finish time. Imagine that a taxi firm has received $n$ journey bookings, each of which has a start time, signifying when the taxi will leave the depot, and a finish time telling us when the taxi is expected to return. How might we assign all of these bookings to vehicles so that the minimum number of vehicles is needed?

Figure 1.5(a) shows an example problem where we have ten taxi bookings. For illustrative purposes these have been ordered from top to bottom according to their start times. It can be seen, for example, that booking 1 overlaps with bookings 2, 3 and 4; hence any taxi carrying out booking 1 will not be able to serve bookings 2, 3 and 4. We can construct a graph from this information by using one vertex for each booking and then adding edges between any vertex pair corresponding to overlapping bookings. A 3-colouring of this example graph is shown in Figure 1.5(b), and the corresponding assignment of the bookings to three taxis (the minimum number possible) is shown in Figure 1.5(c).

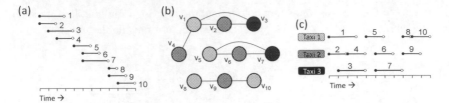

**Fig. 1.5** A set of taxi journey requests over time (a), its corresponding interval graph and 3-colouring (b), and (c) the corresponding assignment of journeys to taxis

In this particular case we see that our example problem has resulted in a graph made of three smaller graphs (components), comprising vertices $v_1$ to $v_4$, $v_5$ to $v_7$ and $v_8$ to $v_{10}$ respectively. However, this will not always be the case and will depend on the nature of the bookings received.

A graph constructed from time-dependent tasks such as this is usually referred to as an *interval graph*. In Chapter 2, we will show that a simple inexpensive algorithm exists for interval graphs that will always produce an optimal solution (that is, a solution using the fewest number of colours possible).

### 1.1.4 Compiler Register Allocation

Our fourth and final example in this section concerns the allocation of computer code variables to registers on a computer processor. When writing code in a particular programming language, whether it be C++, Pascal, FORTRAN or some other option, the programmer is free to make use of as many variables as he or she sees fit. When it comes to compiling this code, however, it is advantageous for the compiler to assign these variables to registers[1] on the processor since accessing and updating values in these locations is far faster than carrying out the same operations using the computer's RAM or cache.

Computer processors only have a limited number of registers. For example, most RISC processors feature 64 registers: 32 for integer values and 32 for floating point values. However, not all variables in a computer program will be in use (or "live") at a particular time. We might therefore choose to assign multiple variables to the same register if they are seen not to interfere with one another.

Figure 1.6(a) shows an example piece of computer code making use of five variables, $v_1, \ldots, v_5$. It also shows the *live ranges* for each variable. So, for example, variable $v_2$ is live only in lines (2) and (3), whereas $v_3$ is live from lines (4) to (9). It can also be seen, for example, that the live ranges of $v_1$ and $v_4$ do not overlap. Hence we might use the same register for storing both of these variables at different periods during execution.

---

[1] Registers can be considered physical parts of a processor that are used for holding small pieces of data.

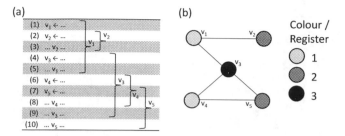

**Fig. 1.6** (a) An example computer program together with the live ranges of each variable. Here, the statement "$v_i \leftarrow \ldots$" denotes the assignment of some value to variable $v_i$, whereas "$\ldots v_i \ldots$" is just some arbitrary operation using $v_i$. (b) shows an optimal colouring of the corresponding interference graph

The problem of deciding how to assign the variables to registers can be modelled as a graph colouring problem by using one vertex for each live range and then adding edges between any pairs of vertices corresponding to overlapping live ranges. Such a graph is known as an *interference graph*, and the task is to now colour the graph using equal or fewer colours than the number of available registers. Figure 1.6(b) shows that in this particular case only three registers are needed: variables $v_1$ and $v_4$ can be assigned to register 1, $v_2$ and $v_5$ to register 2, and $v_3$ to register 3.

Note that in the example of Figure 1.6, the resultant interference graph actually corresponds to an interval graph, rather like the taxi example from the previous subsection. Such graphs will arise in this setting when using straight-line code sequences or when using software pipelining. In most situations however, the flow of a program is likely to be far more complex, involving if-else statements, loops, goto commands, and so on. In these cases the more complicated process of *liveness analysis* will be needed for determining the live ranges of each variable, which could result in an interference graphs of any arbitrary topology (see also Chaitin (2004)).

## 1.2 Why "Colouring"?

We have now seen four practical applications of the graph colouring problem. But why exactly is it concerned with "colouring" vertices? In fact, the graph colouring problem was first noted in 1852 by a student of University College London, Francis Guthrie (1831–1899), who, while colouring a map of the counties of England, noticed that only four colours were needed to ensure that all neighbouring counties were allocated different colours.

To show how the colouring of maps relates to the colouring of vertices in a graph, consider the example map of the historical counties of Wales given in Figure 1.7(a). This particular map involves 16 "regions", including 14 counties, the sea on the left and England bordering on the right. Figure 1.7(d) shows that this map can indeed be

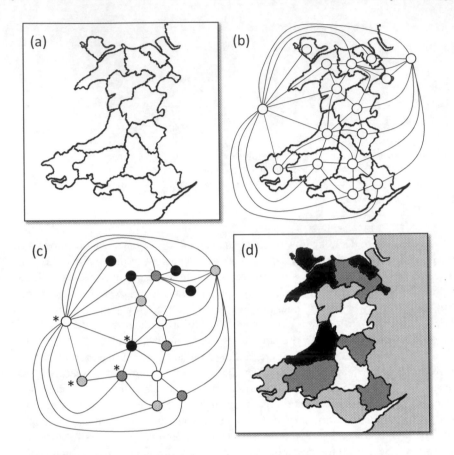

**Fig. 1.7** Illustration of how graphs can be used to colour the regions of a map

coloured using just four colours (light grey, dark grey, black, white). But how does the graph colouring problem itself inform this process?

In Figure 1.7(b), we begin by placing a single vertex in the centre of each region of the map. Next, edges are drawn between any pair of vertices whose regions are seen to share a border. Thus, for example, the vertex appearing in England on the right will have edges drawn to the seven vertices in the seven neighbouring Welsh counties and also to the vertex appearing in the sea on the far left. If we take care in drawing these edges, it can be shown that we will always be able to draw a graph from a map in this way so that no pair of edges needs to cross one another. Technically speaking, a graph that can be drawn with no crossing edges is known as a *planar graph*, of which Figure 1.7(c) is an example.

Figure 1.7(c) also illustrates how we might now colour this planar graph using just four colours. The counties corresponding to these vertices can then be allocated the same colours in the actual map of Wales, as shown in Figure 1.7(d).

We might now ask whether we always need to use exactly four colours to successfully colour a map. In some cases, such as a map depicting a single island region surrounded by sea, less than four colours will obviously be sufficient. On the other hand, for the map of Wales shown in Figure 1.7, we can deduce that exactly four colours will be needed, (a) because a solution using four colours has already been constructed (as shown in the figure), and (b) because a solution using three or fewer colours is impossible. The latter point is due to the fact that the planar graph in Figure 1.7(c) contains a set of four vertices that each have an edge between them (indicated by asterisks in this figure). This obviously tells us that different colours will be needed for each of these vertices.

The fact that, as Francis Guthrie suspected, four colours turn out to be sufficient to colour *any* map (or, equivalently, four colours are sufficient to colour any planar graph) is due to the celebrated Four Colour Theorem, which was ultimately proved in 1976 by Kenneth Appel and Wolfgang Haken of the University of Illinois—a full 124 years after it was first conjectured (see Section 5.1). However, it is important to stress at this point that the Four Colour Theorem does not apply to *all* graphs, but only to *planar* graphs. What can we say about the number of colours that are needed for colouring graphs that are not planar? Unfortunately, as we shall see, in these cases we do not have the luxury of a strong result like the Four Colour Theorem.

## 1.3 Problem Description

We are now in a position to define the graph colouring problem more formally. Let $G = (V, E)$ be a graph, consisting of a set of $n$ vertices $V$ and a set of $m$ edges $E$. Given such a graph, the graph colouring problem seeks to assign each vertex $v \in V$ an integer $c(v) \in \{1, 2, \ldots, k\}$ such that:

- $c(v) \neq c(u) \ \forall \{v, u\} \in E$; and
- $k$ is minimal.

In this interpretation, instead of using actual colours such as grey, black, and white to colour the vertices, we use the labels 1, 2, 3, up to $k$. If we have a solution in which a vertex $v$ is assigned to, say, colour 4, this is then written $c(v) = 4$. According to the first bullet above, pairs of vertices in $G$ that are joined by an edge (usually known as *adjacent* vertices) must be assigned to different colours. The second bullet then states that we are seeking to minimise the number of different colours that are used.

To illustrate these ideas, the example graph depicted in Figure 1.8 has a vertex set $V$ containing $n = 10$ vertices,

$$V = \{v_1, v_2, v_3, v_4, v_5, v_6, v_7, v_8, v_9, v_{10}\},$$

and an edge set $E$ containing $m = 21$ edges,

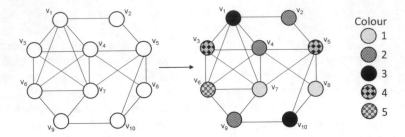

**Fig. 1.8** A small graph with ten vertices and 21 edges (a), and (b) a corresponding 5-colouring

$$E = \{\{v_1,v_2\},\{v_1,v_3\},\{v_1,v_4\},\{v_1,v_6\},\{v_1,v_7\},\{v_2,v_5\},$$
$$\{v_3,v_4\},\{v_3,v_6\},\{v_3,v_7\},\{v_4,v_5\},\{v_4,v_6\},\{v_4,v_7\},$$
$$\{v_4,v_8\},\{v_5,v_7\},\{v_5,v_8\},\{v_5,v_{10}\},\{v_6,v_7\},\{v_6,v_9\},$$
$$\{v_7,v_9\},\{v_8,v_{10}\},\{v_9,v_{10}\}\}.$$

Here, we see that each element of the set $E$ is a pair of vertices $\{v_i, v_j\}$ indicating that these two vertices are adjacent to one another. Note that edges of the form $\{v_i, v_i\}$, commonly referred to as "loops", are not allowed, since this would make it impossible to colour the vertex $v_i$.

Figure 1.8 also shows a 5-colouring of the example graph. Using our notation, this solution can be written as follows:

$$c(v_1) = 3, \; c(v_2) = 2, \; c(v_3) = 4, \; c(v_4) = 2, \; c(v_5) = 4,$$
$$c(v_6) = 5, \; c(v_7) = 1, \; c(v_8) = 1, \; c(v_9) = 2, \; c(v_{10}) = 3.$$

We now give some useful definitions that will help us to describe a graph colouring solution and its properties.

**Definition 1.1** *A colouring of a graph is called* complete *if all vertices* $v \in V$ *are assigned a colour* $c(v) \in \{1,\ldots,k\}$*; else the colouring is considered* partial.

**Definition 1.2** *A* clash *describes a situation where a pair of adjacent vertices* $u, v \in V$ *are assigned the same colour (that is,* $\{u,v\} \in E$ *and* $c(v) = c(u)$*). If a colouring contains no clashes, then it is considered* proper*; else it is considered* improper.

**Definition 1.3** *A colouring is* feasible *if and only if it is both complete and proper.*

**Definition 1.4** *The* chromatic number *of a graph G, denoted by* $\chi(G)$*, is the minimum number of colours required in a feasible colouring of G. A feasible colouring of G using exactly* $\chi(G)$ *colours is considered* optimal.

For example, the 5-colouring shown in Figure 1.8 is feasible because it is both complete (all vertices have been allocated colours) and proper (it contains no clashes). In this case the chromatic number of this graph $\chi(G)$ is already known to be 5, so the colouring can also be said to be optimal.

Some further useful definitions for the graph colouring problem involve colour classes, and structures known as *cliques* and *independent sets*.

**Definition 1.5** *A* colour class *is a set containing all vertices in a graph that are assigned to a particular colour in a solution. That is, given a particular colour $i \in \{1,\dots,k\}$, a colour class is defined as the set $\{v \in V : c(v) = i\}$.*

**Definition 1.6** *An* independent set *is a subset of vertices $I \subseteq V$ that are mutually nonadjacent. That is, $\forall u, v \in I, \{u, v\} \notin E$.*

**Definition 1.7** *A* clique *is a subset of vertices $C \subseteq V$ that are mutually adjacent: $\forall u, v \in C, \{u, v\} \in E$.*

To illustrate these definitions, two example colour classes from Figure 1.8 are $\{v_2, v_4, v_9\}$ and $\{v_8\}$. Example independent sets from this figure include $\{v_2, v_7, v_8\}$ and $\{v_3, v_5, v_9\}$. The largest clique in Figure 1.8 is $\{v_1, v_3, v_4, v_6, v_7\}$, though numerous smaller cliques also exist, such as $\{v_6, v_7, v_9\}$ and $\{v_2, v_5\}$.

Given the above definitions, it is also useful to view graph colouring as a type of partitioning problem where a solution $S$ is represented by a set of $k$ colour classes $S = \{S_1, \dots, S_k\}$. In order for $S$ to be feasible it is then necessary that the following constraints be obeyed:

$$\bigcup_{i=1}^{k} S_i = V, \tag{1.1}$$

$$S_i \cap S_j = \emptyset \quad (1 \le i \ne j \le k), \tag{1.2}$$

$$\forall u, v \in S_i, \{u, v\} \notin E \ (1 \le i \le k), \tag{1.3}$$

with $k$ being minimised. Here, Constraints (1.1) and (1.2) state that $S$ should be a partition of the vertex set $V$ (that is, all vertices should be assigned to exactly one colour class each). Constraint (1.3) then stipulates that no pair of adjacent vertices should be assigned to the same colour class (i.e., all colour classes in the solution should be independent sets). Referring to Figure 1.8 once more, the depicted solution using this interpretation of the problem is now written $S = \{\{v_1, v_{10}\}, \{v_2, v_4, v_9\}, \{v_3, v_5\}, \{v_6\}, \{v_7, v_8\}\}$.

## 1.4 Problem Complexity

Now that we have properly defined the graph colouring problem, the natural question to ask is: "what algorithm can be employed to solve it?" In this question we use the word "solve" in the strong sense—that is, an algorithm is said to solve the graph colouring problem if it is able to take *any* graph (of any size and any topology) and return an optimal solution in all cases. Is such an algorithm achievable?

## 1.4.1 Solution Space Growth Rates

One possible, though perhaps foolhardy, method for solving the graph colouring problem is to check every possible assignment of vertices to colours and then return the best of these (that is, return the solution from the set of all possible assignments that is both feasible and seen to be using the fewest colours). Such a method would indeed be guaranteed to return an optimal solution, but would it work in practice? Following this approach, given a particular graph with $n$ vertices, we would first need to decide the maximum number of colours that the solution might use. In practice, this could be estimated as some value between 1 and $n$, but for now we will assume this value to simply be $n$, since no feasible solution will ever require more colours than vertices.

Consider now the number of assignments that would need to be checked. Since there would be $n$ choices of colour for each of the $n$ vertices, this would give a solution space containing a total of $n^n$ candidate solutions. This number obviously grows very rapidly with regard to $n$, meaning that the number of candidate solutions to be checked will quickly become too large for even the most powerful computer to tackle. To illustrate, a graph with $n = 50$ vertices would lead to over $50^{50} \approx 8.8 \times 10^{84}$ different assignments: a truly astronomical number. This would make the task of creating and checking all of these assignments, even for this modestly sized problem, far beyond the computing power of all of the world's computers combined. (For comparison's sake, the number of atoms in the known universe is thought to be around $10^{82}$.) In addition, even if we were to limit ourselves to candidate solutions using a maximum of $k < n$ colours labelled $1, \ldots, k$, this would still lead to $k^n$ assignments needing to be considered, a function that is still subject to a similar combinatorial explosion for any $k > 1$.

A slightly better, though still ultimately doomed approach would be to make note of the symmetry that exists in the assignment method just described. It is worth noting at this point that when we allocate "colours" to vertices, these are essentially arbitrary labels, and what we are more interested in is the *number of different colours* being used as opposed to what these labels actually are. To these ends, a solution such as

$$c(v_1) = 3, \; c(v_2) = 2, \; c(v_3) = 4, \; c(v_4) = 2, \; c(v_5) = 4,$$
$$c(v_6) = 5, \; c(v_7) = 1, \; c(v_8) = 1, \; c(v_9) = 2, \; c(v_{10}) = 3$$

can be considered identical to the solution

$$c(v_1) = 1, \; c(v_2) = 2, \; c(v_3) = 4, \; c(v_4) = 2, \; c(v_5) = 4,$$
$$c(v_6) = 5, \; c(v_7) = 3, \; c(v_8) = 3, \; c(v_9) = 2, \; c(v_{10}) = 1,$$

because the makeup of the five colour classes in both cases are equivalent, with only the labels of the being different (in this example, colours 1 and 3 have been swapped). Stated more precisely, this means that if we are given a candidate solution using $k$ different colours with labels $1, 2, \ldots, k$, the number of different assignments that would actually specify the same solution would be $k! = k \times (k-1) \times (k-2) \times \ldots \times 2 \times 1$, representing all possible assignments of the $k$ labels to the colour classes. The space of all possible assignments of vertices to colours is therefore far larger (exponentially so) than it needs to be.

Perhaps a better approach than considering all possible assignments is to therefore turn to the alternative but equivalent statement of the graph colouring problem from Section 1.3. Here we seek to partition the vertex set into a set of $k$ colour classes $S = \{S_1, \ldots, S_k\}$, where each set $S_i \in S$ is an independent set, with $k$ being minimised. The number of ways of partitioning a set with $n$ items into nonempty subsets is given by the $n$th Bell number, denoted by $B_n$. For example, the third bell number $B_3 = 5$ because a set with three elements $v_1$, $v_2$, and $v_3$, can be partitioned in five separate ways:

$$\{\{v_1, v_2, v_3\}\},$$
$$\{\{v_1\}, \{v_2\}, \{v_3\}\},$$
$$\{\{v_1\}, \{v_2, v_3\}\},$$
$$\{\{v_2\}, \{v_1, v_3\}\}, \text{and}$$
$$\{\{v_3\}, \{v_1, v_2\}\}.$$

In contrast to the previous case where up to $n^n$ assignments need to be considered, the use of sets here means that the *labelling* of the colour classes is not now relevant, meaning that the issues surrounding solution symmetry have been resolved. This also has the effect of reducing the size of the solution space considerably. A suitable approach for colouring a graph with $n$ vertices might now simply involve enumerating all $B_n$ possible partitions and simply checking for the one that comprises the fewest independent sets. Further improvements could also be achieved if we were to limit the number of available colours to some value $k < n$, reducing the size of the solution space even further. Stirling numbers of the second kind, commonly denoted by $\{{n \atop k}\}$, define the number of ways of partitioning $n$ items into exactly $k$ nonempty subsets, and can be calculated by the formula:

$$\begin{Bmatrix} n \\ k \end{Bmatrix} = \frac{1}{k!} \sum_{j=0}^{k} (-1)^{k-j} \binom{k}{j} j^n. \tag{1.4}$$

So, for instance, the number of ways of partitioning three items into exactly two nonempty subsets is $\{{3 \atop 2}\} = 3$, because we have three different options:

$$\{\{v_1\}, \{v_2, v_3\}\},$$
$$\{\{v_2\}, \{v_1, v_3\}\}, \text{and}$$
$$\{\{v_3\}, \{v_1, v_2\}\}.$$

Note that summing Stirling numbers of the second kind for all values of $k$ from 1 to $n$ leads to the $n$th Bell number:

$$B_n = \sum_{k=1}^{n} \left\{ {n \atop k} \right\}. \tag{1.5}$$

We might now choose to employ an enumeration algorithm that starts by considering $k = 1$ colour. At each step the algorithm then simply needs to check all $\left\{ {n \atop k} \right\}$ possible partitions to see if any correspond to a feasible solution ($k$-colouring). If such a solution is found, the algorithm can halt immediately with the knowledge that an optimal solution has been found. Otherwise, $k$ would need to be incremented by 1, with the process continuing as before. Ultimately such an algorithm would therefore need to consider a maximum of

$$\sum_{k=1}^{\chi(G)} \left\{ {n \atop k} \right\} \tag{1.6}$$

candidate solutions. But is such an approach suitable for solving the graph colouring problem? Unfortunately not. Even though our original solution space size of $n^n$ has been reduced quite considerably by representing solutions as partitions, Stirling numbers of the second kind still exhibit exponential growth rates for most values of $k$. As an example, if we were seeking to produce and examine all partitions of 50 items into 10 subsets, which is again quite a modestly sized graph colouring problem, this would lead to $\left\{ {50 \atop 10} \right\} \approx 2.6 \times 10^{43}$ candidate solutions. Such a figure is still well beyond the reach of any contemporary computing resources.

### 1.4.2 Problem Intractability

The above discussions have demonstrated that any graph colouring algorithm based on enumerating and checking the entire solution space is not sensible because, on anything except trivial problem instances, its execution will simply take too long. However, the exponential growth rate of the solution space is not the sole reason why the graph colouring problem is so troublesome, since many "easy to solve" problems also feature similarly large solution spaces. As an example, consider the computational problem of sorting a collection of integers into ascending order. Given a set of $n$ unique integers, there are a total of $n!$ different ways of arranging these, and only one of these candidate solutions will give us the required answer. However, it would be completely unnecessary to employ an algorithm that went about checking all possible $n!$ candidate solutions, because a multitude of efficient algorithms whose complexities are not subject to a combinatorial explosion are available for the sorting problem, including the QUICKSORT and MERGESORT algorithms. Detailed descriptions of these can be found in the comprehensive textbook of Cormen et al. (2009), amongst other places.

In contrast to the problem of sorting, the fact that a problem like graph colouring can be considered "hard", or "intractable", is due to the pioneering work of Stephen Cook in the early 1970s (Cook, 1971). In this work Cook introduced the concepts of

NP-completeness and polynomial-time reductions, and also showed that a problem known as the "satisfiability problem" is NP-complete.

In short, given a particular Boolean expression, the satisfiability problem asks whether there exists some assignment of values to the variables such that the expression evaluates to "true". For example, the Boolean expression $(x_1 \land \neg x_2)$ is satisfiable because an assignment of, say, true to $x_1$ and false to $x_2$ leads to $(T \land \neg F) = (T \land T) = T$ as required. On the other hand the expression $(x_1 \land \neg x_1)$ is not satisfiable since all possible assignments to the variables evaluate to a false: $(T \land \neg T) = (T \land F) = F$, and $(F \land \neg F) = (F \land T) = F$. Cook proved that there is no known algorithm for efficiently solving all instances of the satisfiability problem, labelling this characteristic as NP-complete. It is this finding that leads to the conclusion that graph colouring is also an NP-complete problem, as we shall now see.

In the field of computational complexity, problems are typically posed as "decision problems" requiring "yes" or "no" answers. Graph colouring problems can easily be stated as decision problems by simply posing the problem in the form "Can a graph $G$ be coloured feasibly using $k$ colours?", as opposed to asking what the minimum number of colours actually is. Decision problems for which there exist "good" algorithms—that is, algorithms whose growth rates (and therefore execution times) can be expressed via polynomials and are therefore not subject to a combinatorial explosion—are said to belong to the class P, standing for polynomial-time problems.

Another class of decision problems is NP, standing for nondeterministic polynomial time. A problem is said to be a member of NP if, given a particular candidate solution together with a claim that it gives a "yes" answer, there exists a polynomial-time algorithm for verifying whether this is the case. For example, the decision variant of the graph colouring problem ("Can $G$ be feasibly coloured using $k$ colours?") belongs to NP because, given any candidate solution and value for $k$, it is easy to check in polynomial time whether it is both feasible and using exactly $k$ colours.

Clearly any problem belonging to P also belongs to NP, since for any problem in P, we can simply use its available polynomial-time algorithm to solve (and therefore verify) it. This means that P $\subseteq$ NP. However, despite decades of research, it is not known whether in fact P = NP. Indeed, it is generally believed across the academic fields that the opposite is true and that P $\neq$ NP. This brings us to the class of NP-complete problems. Problems that are NP-complete belong to the set NP and can therefore have potential solutions verified in polynomial time. However, an algorithm for locating a "yes" solution to such problems in the first place will still have to resort to enumerating and checking a significant portion of the solution space. Since NP-complete problems feature solution spaces that are known to grow exponentially in relation to problem size, this implies that there is no known polynomially bounded algorithm for achieving this task.

So how do we know that the decision variant of the graph colouring problem is NP-complete? As mentioned, Cook showed in 1971 that the satisfiability problem is NP-complete, proving the existence of at least one NP-complete problem. Fortunately, however, this result can be used to prove the NP-completeness of a whole host of other problems via the use of what are known as polynomial transformations. A polynomial transformation is essentially a way of efficiently transforming one decision problem into another. Let $D_1$ and $D_2$ be two separate decision problems. We say that $D_1$ can be polynomially transformed into $D_2$, written $D_1 \propto D_2$, if there is a polynomial-time method of transforming any instance of $D_1$ into an instance of $D_2$, such that the answer to $D_1$ is "yes" if and only if the corresponding answer to $D_2$ is also "yes". Another way of looking at this is to say that if $D_1 \propto D_2$, then $D_2$ *generalises* $D_1$ in that all instances of $D_1$ can be converted into corresponding instances of $D_2$ (though $D_2$ might feature other instances that do not have corresponding instances in $D_1$). The reason why polynomial transformations are useful is that, if $D_1$ is known to be NP-complete and $D_1 \propto D_2$, then this also proves that $D_2$ is NP-complete, since all possible instances of $D_1$ can be transformed and stated as a corresponding problem in $D_2$. (If, for purposes of contradiction, $D_1$ was known to be NP-complete, and $D_1 \propto D_2$, but $D_2$ was known to be in P, this would imply that all instances of $D_1$ could also be solved in polynomial time, since we could just use a polynomial transformation process to convert any instance of $D_1$ into $D_2$ and then solve this using an efficient (polynomially bounded) algorithm. This would contradict the initial statement that $D_1$ was NP-complete.)

The decision variant of graph colouring is known to be NP-complete because it has been shown to generalise the NP-complete problem of 3-satisfiability (i.e., the 3-satisfiability problem is polynomially reducible to the graph colouring problem). In turn, the 3-satisfiability problem generalises the satisfiability problem itself, which we have already established as being NP-complete due to the findings of Cook (1971).

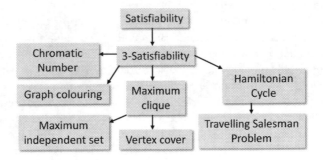

**Fig. 1.9** Some paths of polynomial reductions

The tree in Figure 1.9 depicts this chain of reductions, along with some other famous examples of NP-complete problems. Related to graph colouring, these problems include:

- The chromatic number problem: Given a graph $G$ and some integer $k$, is $G$'s chromatic number $\chi(G) = k$?
- The maximum clique problem: Given a graph $G$ and some integer $k$, is there a clique in $G$ containing $k$ vertices?

- The maximum independent set problem: Given a graph $G$ and some integer $k$, is there an independent set in $G$ containing $k$ vertices?

Proofs of many of these reductions can be found in the seminal paper of Karp (1972). A number of similar proofs are also detailed in Cormen et al. (2009).

When talking about the graph colouring problem, or indeed many other computationally intractable problems, it is also common to see the term "NP-hard" instead of NP-complete. NP-hard problems are at least as hard as NP-complete problems because they are not required to be in the class NP (i.e., they do not have to be stated as decision problems). Consequently, NP-hard problems will often be stated as an optimisation version of a corresponding NP-complete decision problem. So with graph colouring for instance, rather than asking "is there a feasible colouring of $G$ that uses $k$ colours?" (for which the answer will be "yes" or "no"), the problem will now be stated: "how might we colour the vertices of $G$ such that the minimum number of different colours is being used?" Note that the non-decision variant version of the graph colouring algorithm as stated in Section 1.3 is itself NP-hard, as are the corresponding optimisation versions of the chromatic number, maximum clique and maximum independent set problems.

Rounding off this section, we should note that the existence of NP-complete and NP-hard problems is based wholly on the conjecture that P $\neq$ NP. The question of whether this conjecture is actually true is one of the most famous unsolved problems in the mathematical sciences. Indeed, the Clay Mathematics Institute has offered a prize of one million US dollars to anyone who is able to offer a formal proof one way or the other. As we have mentioned, it is largely believed that P $\neq$ NP, but if the opposite were found to be true, it would have hugely profound scientific consequences, not least because it would lead to the availability of polynomial-time algorithms for large numbers of computational problems previously thought to be intractable, including graph colouring itself. The reader is referred to the excellent textbook of Garey and Johnson (1979) for a full introduction to this topic.

## 1.5 Can We Solve the Graph Colouring Problem?

In the previous section we arrived at the inconvenient conclusion that, as with all NP-hard problems, an efficient (polynomial-time) algorithm for solving the graph colouring problem is almost certainly beyond our grasp. In this context we have again used the word "solve" in the strong sense—that is, an algorithm "solves" a particular problem if and only if it returns the optimal solution for all problem instances. For the colouring of any arbitrary graph, we will therefore need to turn towards approximation algorithms and/or heuristic methods

Approximation algorithms and heuristic methods are used for achieving approximate solutions to intractable problems and typically operate in polynomial time. They do not "solve" their intended problem, but they do attempt to provide solutions that are hopefully acceptable for the user. Considering the graph colouring problem, we might choose to employ a particular algorithm $A$ on graph $G$. After

a short amount of time, $A$ might then return a feasible $k$-colouring for $G$. If a solution using $k$ colours is acceptable for use, we might stop there; however, if it is
unsuitable, we may then choose to modify the behaviour of $A$ or indeed try another
algorithm $B$ which will hopefully give us better solution, perhaps using fewer than
$k$ colours. This sort of process highlights the underlying issue with approximation
algorithms and heuristics in that they will often return suboptimal solutions. Indeed,
even if they *do* happen to return the optimal solution, we may not be able to prove
this fact. Further, they may happen to produce quite good solutions in comparison
to other algorithms on some types of graph, but poor solutions on others.

The fact that graph colouring is NP-hard does not mean that *no* problem instances are solvable in polynomial time, however. As a trivial example, consider a
graph comprising a vertex set $V = \{v_1,\dots,v_n\}$ and edge set $E = \emptyset$ (such a graph
is commonly known as an *empty graph on n vertices*). Obviously, since no pairs of
vertices are adjacent in this graph, the $n$ vertices can all be feasibly assigned the
same single colour, giving a chromatic number of 1. In practice it would be easy
to write an algorithm to check whether $E = \emptyset$ and, if this is the case, produce the
corresponding optimal solution.

In the following subsections we will now take a look at a selection of some less
trivial graph topologies for which exact results on the chromatic number are known.
In Chapter 2, we will also see two heuristic algorithms for graph colouring that, in
addition to producing good results on arbitrary graphs, also turn out to be exact for
some of these examples.

### 1.5.1 Complete Graphs

Complete graphs with $n$ vertices, denoted by $K_n$, are graphs that feature an edge
between each pair of vertices, giving a set $E$ of $m = \frac{n(n-1)}{2}$ edges. It is obvious that
because all vertices in the complete graph are mutually adjacent, all vertices must be
assigned to their own individual colour. Hence the chromatic number of a complete
graph $\chi(K_n) = n$. Example optimal solutions for $K_1$ to $K_5$ are shown in Figure 1.10.

**Fig. 1.10** Optimal colourings of the complete graphs (from left to right) $K_1$, $K_2$, $K_3$, $K_4$ and $K_5$

## 1.5.2 Bipartite Graphs

Bipartite graphs, denoted by $G = (V_1, V_2, E)$, are graphs whose vertices can be partitioned into two sets $V_1$ and $V_2$ such that edges only exist between vertices in $V_1$ and vertices in $V_2$. As a result $V_1$ and $V_2$ are both independent sets, meaning that bipartite graphs can be coloured using just two colours, with all vertices in $V_1$ being assigned to one colour, and all vertices in $V_2$ being assigned to the other. It is obvious, therefore, that a graph $G$ features a chromatic number $\chi(G) = 2$ if and only if it is bipartite.

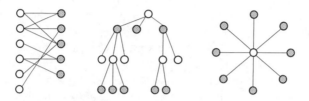

**Fig. 1.11** Optimal colourings of (from left to right) an arbitrary bipartite graph, a tree, and a star graph

Figure 1.11, shows three examples of bipartite graphs and their optimal colourings: an arbitrary bipartite graph; a tree (that is, a bipartite graph that contains no cycles); and a star graph.

## 1.5.3 Cycle, Wheel and Planar Graphs

Cycle graphs, denoted by $C_n$, where $n \geq 3$, comprise a set of vertices $V = \{v_1, \ldots, v_n\}$ and a set of edges $E$ of the form $\{\{v_1, v_2\}, \{v_2, v_3\}, \ldots, \{v_{n-1}, v_n\}, \{v_n, v_1\}\}$. The cycle graphs $C_3, C_4, C_8$, and $C_9$ are shown in Figure 1.12(a).

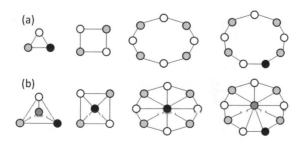

**Fig. 1.12** Optimal colourings of (a) cycle graphs $C_3, C_4, C_8$, and $C_9$, and (b) wheel graphs $W_4, W_5$, $W_9$, and $W_{10}$

It is known that only two colours are needed to colour $C_n$ when $n$ is even (an even cycle); hence even cycles are a type of bipartite graph. However, three colours are needed when $n$ is odd (an odd cycle). This is illustrated in the figure, where $\chi(C_4) = \chi(C_8) = 2$ whereas $\chi(C_3) = \chi(C_9) = 3$.

To explain this result, first consider the even cycle case. To construct a 2-colouring we simply choose an arbitrary vertex and colour it white. We then proceed around the graph in a clockwise direction colouring the second vertex grey, the third vertex white, the fourth grey, and so on. When we reach the $n$th vertex, this can be coloured grey because the two vertices adjacent to it, namely the first and $(n-1)$th vertex will both be coloured white. Hence only two colours are required.

On the other hand, when $n$ is odd (and $n \geq 3$), three colours will be required. Following the same pattern as the even case, an initial vertex is chosen and coloured white, with other vertices in a clockwise direction being assigned grey, white, grey, white, as before. However, when the $n$th vertex is reached, this will be adjacent to both the $(n-1)$th vertex (coloured grey), and the first vertex (coloured white). Hence a third colour will be required to feasibly colour the graph.

Wheel graphs with $n$ vertices, denoted by $W_n$, are obtained from the cycle graph $C_{n-1}$ by adding a single extra vertex $v_n$ together with the additional edges $\{v_1, v_n\}, \{v_2, v_n\}, \ldots, \{v_{n-1}, v_n\}$. Example wheel graphs are shown in Figure 1.12(b). It is clear that similar results to cycle graphs can be stated for wheel graphs. Specifically, when $n$ is odd, three colours will be required to colour $W_n$ because the graph will be composed of the even cycle $C_{n-1}$, requiring two colours, and the additional vertex $v_n$ which, being adjacent to all vertices in $C_{n-1}$, will require a third colour. Similarly, when $n$ is even, $\chi(W_n) = 4$ because the graph will be composed of the odd cycle $C_{n-1}$, requiring three colours, together with vertex $v_n$, which will require a forth colour.

It is clear from the illustrations in Figure 1.12 that cycle graphs and wheel graphs (of any size) are both particular cases of planar graphs, in that they can be drawn on a two-dimensional plane without any of the edges crossing. This fits in with the Four Colour Theorem, which states that if a graph is planar then it can be feasibly coloured using four or fewer colours (see Sections 1.2 and 5.1). However, the Four Colour Theorem does not imply that if a graph is 4-colourable then it must also be planar, as the next example illustrates.

### 1.5.4 Grid Graphs

Grid graphs can be formed by placing all vertices in a lattice formation on a two-dimensional plane. In a *sparse* grid graph, each vertex is adjacent to four vertices: the vertex above it, the vertex below it, the vertex to the right, and the vertex to the left (see Figure 1.13(a)). For a *dense* grid graph, a similar pattern is used, but vertices are also adjacent to vertices on their surrounding diagonals (Figure 1.13(b)).

A practical application of such graphs occurs in the arrangement of seats in exam venues. Imagine a large examination venue where the desks have been placed in a

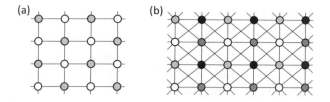

**Fig. 1.13** Optimal colourings of (a) a sparse grid graph and (b) a dense grid graph

grid formation. In such cases we might want to avoid instances of students copying from each other by making sure that each student is always seated next to students taking different exams. What is the minimum number of exams that can take place in the venue if this is the case? This problem can be posed as a graph colouring problem by representing each desk as a vertex, with edges representing pairs of desks that are close enough for students to copy from.

If it is assumed that students can only copy from students seated in front, behind, to their left, or to their right, we get the sparse grid graph shown in Figure 1.13(a). Though perhaps not obvious by inspection, this graph is a type of bipartite graph since it can be coloured using just two colours according to the pattern shown. Hence a minimum of two exams can take place in this venue at any one time.

In circumstances where students are able to copy from students sitting on any of the eight desks surrounding them, we get the dense grid graph shown in Figure 1.13(b). As illustrated, this grid can be coloured using four colours according to the pattern shown. In this graph each vertex, together with the vertex above, the vertex on the right, and the vertex on the upper diagonal right, forms a clique of size four. Hence we can conclude that a feasible colouring using fewer than four colours does not exist.

The dense grid graph also provides a simple example of a graph that is nonplanar but is still 4-colourable. Although cliques of size 4 are themselves planar, the nature by which the various cliques interlock in this example means that some edges will always need to cross one another. This illustrates that a graph does not have to be planar in order for it to be colourable with four or fewer colours.

## 1.6 About This Book

As we have seen in this introductory chapter, this book focusses on the problem of colouring the *vertices* of a graph. Sometimes the term "graph colouring" is also applied to the task of colouring the edges of a graph or the faces of a graph. However, as we shall see in Chapter 5, edge and face colouring problems can easily be transformed into an equivalent vertex colouring problem via the concepts of line graphs and dual graphs respectively. Consequently, unless explicitly stated otherwise, the term "graph colouring" in this book refers exclusively to the problem of vertex colouring.

This book is aimed primarily at the algorithmic and heuristic aspects of graph colouring. That is, rather than providing in-depth treatments of the large number of theorems and conjectures surrounding certain graph topologies, we choose to focus on the characteristics of different algorithms for the general graph colouring problem. Do these algorithms provide optimal solutions to some graphs? How do they perform on various different topologies where the chromatic number is unknown? Why are some algorithms better for some types of graphs, but worse for others? What are run time characteristics of these algorithms?

In addition, this book also examines many of the real-world operational research problems that can be tackled using graph colouring techniques. As we will see, these include problems as diverse as the colouring of maps, the production of round-robin tournaments, Sudoku, assigning variables to computer registers, and checking for short circuits on printed circuit boards. Individual chapters are also allocated to in-depth examinations of the problems of designing seating plans for weddings, scheduling fixtures for sports leagues, and timetabling lectures at universities.

### 1.6.1 Algorithm Implementations

This book is accompanied by a suite of nine graph colouring algorithm implementations, which can be downloaded from:

> http://rhydlewis.eu/resources/gCol.zip

Each of the algorithms included in this suite will be described and analysed in detail in this book. Specifically, the GREEDY, RLF, and DSATUR algorithms are discussed in Chapter 2, and the TABUCOL, PARTIALCOL, hybrid evolutionary algorithm, ANTCOL, hill-climbing, and backtracking algorithms are described in Chapter 4. All of these implementations are written in the C++ programming language and can be executed from the command line using common input and output protocols. A user manual and compilation instructions are provided in Appendix A. Readers are encouraged to make use of these algorithms on their own graph colouring instances and are also invited to modify the code in any way they see fit.

As we shall see, when gauging the effectiveness of a graph colouring algorithm (or any algorithm for that matter) it is important to consider the amount of computational effort required to produce a solution of a given quality. Ideally, we should try to steer clear of measures such as wall-clock time or CPU time because these are largely influenced by the chosen hardware, operating systems, programming languages and compiler options. (That said, CPU and wall-clock time are still measures that are, perhaps regrettably, widely used in the graph colouring literature.)

A more rigorous approach to measuring computational effort involves examining the number of *atomic operations* performed by an algorithm during execution. For classical computational problems such as searching through or sorting the elements of a vector, these are usually considered to be the constant-time operations of comparing two elements and swapping two elements. For others, such as the bin packing problem, where we seek to partition a set of weighted items into a set of bins with limited capacity, these are often considered to be the operations of looking up some feature of the problem instance, such as referencing the weight of one of the items.

For graph colouring algorithms, it is useful to follow this scheme by gauging computational effort via the number of *constraint checks* that are performed. Essentially, a constraint check is considered to occur whenever an algorithm requests some information about a graph, such as whether two vertices are adjacent or not. We will define these operations presently, though it is first necessary to describe how graphs are to be stored in computer memory.

### 1.6.1.1  Representing Graphs

We saw in Section 1.3 that a graph $G$ can be defined by a set $V$ of $n$ vertices and a set $E$ of $m$ edges. While the use of sets in this way is mathematically convenient, implementations of graph algorithms, including our own, usually make use of two different structures, namely *adjacency lists* and *adjacency matrices*. We now define these.

**Definition 1.8**  *Given a graph $G = (V,E)$, an* adjacency list *is a vector of length n, where each element $Adj_v$ corresponds to a list containing all vertices adjacent to vertex v, for all $v \in V$.*

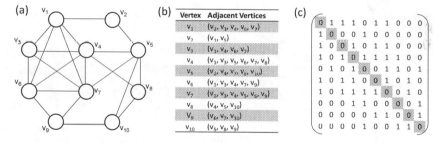

**Fig. 1.14**  A ten-vertex graph (a), its adjacency list (b), and its adjacency matrix (c)

An example adjacency list for a ten-vertex graph is shown in Figure 1.14. The length of the list, $|Adj_v|$, tells us the number of vertices that are adjacent to a vertex $v$. This is usually known as the *degree* of a vertex (see Definition 2.2). So, for example, vertex $v_2$ in this graph is seen to be adjacent to vertices $v_1$ and $v_5$, and therefore has a degree of 2. Note that the sum of all list lengths $\sum_{\forall v \in V} |Adj_v| = 2m$ since, if $v$ appears in a vertex $u$'s adjacency list, then $u$ will also appear in $v$'s adjacency list.

In algorithm implementations, adjacency lists are useful when we are interested in identifying (and presumably doing something to) all vertices adjacent to a particular vertex $v$. On the other hand, they are less useful when we want to quickly answer the question "are vertices $u$ and $v$ adjacent?", as to do so would require searching through either $Adj_u$ or $Adj_v$. For these situations it is therefore preferable to use an adjacency matrix.

**Definition 1.9** *Given a graph* $G = (V, E)$, *an* adjacency matrix *is a matrix* $\mathbf{A}_{n \times n}$ *for which* $A_{ij} = 1$ *if and only if vertices* $v_i$ *and* $v_j$ *are adjacent, and* $A_{ij} = 0$ *otherwise.*

An example adjacency matrix is also provided in Figure 1.14. When considering graph colouring problems, note that edges are not directed, and graphs cannot contain loops. Consequently $\mathbf{A}$ is symmetric ($A_{ij} = A_{ji}$) and only has 0s along its main diagonal ($A_{ii} = 0$).

When implemented, adjacency matrices require memory for storing $n^2$ pieces of information, regardless of the number of edges in the graph. Consequently, they are sometimes considered quite unwieldy, particularly for sparse graphs. However, for modern-day graph colouring algorithms, including those contained in our suite, it is beneficial to make use of both adjacency lists *and* adjacency matrices. This allows for increased speed at the cost of increased memory requirements.

Finally, as we shall see in Chapter 4, our graph colouring algorithms will often attempt to improve solutions by changing the colours of certain vertices. In our implementations it is therefore also useful to make use of an additional matrix $\mathbf{C}_{n \times k}$ for representing graphs where, given a particular $k$-coloured solution $S = \{S_1, ..., S_k\}$, the element $C_{vj}$ gives the number of vertices in colour class $S_j$ that are adjacent to vertex $v$. Full descriptions of how this matrix is used are given in Sections 4.1.1 and 4.1.2.

### 1.6.1.2 Measuring Computational Effort

Having specified the way in which graphs are stored by our algorithm implementations, we are now in a position to define how constraint checks are counted:

1. The task of checking whether two vertices $u$ and $v$ are adjacent is performed using the adjacency matrix $\mathbf{A}$. Accessing element $A_{uv}$ counts as one constraint check.
2. The task of going through all vertices adjacent to a vertex $v$ involves accessing all elements of the list $Adj_v$. This counts as $|Adj_v|$ constraint checks.
3. Determining the degree of a vertex $v$ involves looking up the value $|Adj_v|$. This counts as one constraint check.
4. Determining the number of vertices in colour class $S_i \in S$ that are adjacent to a particular vertex $v$ involves accessing element $C_{vi}$. This counts as one constraint check.

## 1.7 Chapter Summary

In this chapter we have introduced the graph colouring problem, provided a number of examples of its practical applications, and considered the reasons as to why, in the general case, graph colouring should be considered "intractable". In the next chapter we will seek to gain deeper insights into this problem by analysing and comparing three well-known heuristic-based constructive algorithms for this problem. We will also discuss some of the upper and lower bounds that can be established on the chromatic number of a graph, both for the general case and for special types of topologies.

# Chapter 2
# Bounds and Constructive Algorithms

Towards the end of Chap. 1 we saw a variety of different types of graphs that are relatively straightforward to colour optimally, including complete graphs, bipartite graphs, cycle and wheel graphs, and grid graphs. With regard to the chromatic number, we also saw that it is easy to determine when $\chi(G) = 1$ ($G$ is an empty graph), and when $\chi(G) = 2$ ($G$ is bipartite). But can we go further than this? In this chapter we review and analyse a number of fast constructive algorithms for the graph colouring problem. We also make statements on how we are able to bound the chromatic number.

The fact that graph colouring is an intractable problem implies that there is a limited amount that we can say about the chromatic number of an arbitrary graph in general. One simple rule is that, given a graph $G$ with $n$ vertices and $m$ edges, if $m > \lfloor n^2/4 \rfloor$ then $\chi(G) \geq 3$, since any graph fitting this criteria must contain a triangle and therefore cannot be bipartite (Bollobás, 1998); however, even the problem of deciding whether $\chi(G) = 3$ is NP-complete for arbitrary graphs.

In spite of this rather bleak situation, a variety of heuristic-based approximation algorithms are available for graph colouring that are often able to produce very pleasing results. In this chapter we will consider three fast constructive methods which operate by assigning each vertex to a colour one at a time using rules that are intended to keep the overall number of colours as small as possible. As we will see, for certain graph topologies some of these algorithms turn out to be exact, though in most cases they only produce approximate solutions. The first of these algorithms, the so-called GREEDY algorithm, is perhaps the most fundamental method in the field of graph colouring and is also useful for establishing bounds on the chromatic number. Towards the end of the chapter we also present an empirical comparison of the three constructive algorithms in order to provide information on their relative strengths and weaknesses.

At this point it is useful to introduce some further graph terminology. Recall that a graph $G = (V,E)$ is defined by a vertex set $V$ of $n$ vertices and an edge set $E$ of $m$ edges.

**Definition 2.1** *If $\{u,v\} \in E$, vertices $u$ and $v$ are said to be* adjacent. *Vertices $v$ and $u$ are also said to be* incident *to the edge $\{u,v\} \in E$. If $\{u,v\} \notin E$, then vertices $u$ and $v$ are* nonadjacent.

© Springer International Publishing Switzerland 2016
R.M.R. Lewis, *A Guide to Graph Colouring*,
DOI 10.1007/978-3-319-25730-3_2

**Definition 2.2** *The* neighbourhood *of a vertex v, written* $\Gamma_G(v)$*, is the set of vertices adjacent to v in the graph G. That is,* $\Gamma_G(v) = \{u \in V : \{v,u\} \in E\}$*. The* degree *of a vertex v is the cardinality of its neighbourhood set,* $|\Gamma_G(v)|$*, usually written* $\deg_G(v)$*. When the graph being referred to is made clear by the text, these can be written in their shorter forms,* $\Gamma(v)$ *and* $\deg(v)$*, respectively.*

**Definition 2.3** *The* density *of a graph* $G = (V,E)$ *is the ratio of the number of edges to the number of pairs of vertices. For a simple graph with no loops this is calculated* $m/((n(n-1))/2)$*. Graphs with low densities are often referred to as* sparse *graphs; those with high densities are known as* dense *graphs.*

**Definition 2.4** *A graph* $G' = (V',E')$ *is a* subgraph *of G, denoted by* $G' \subseteq G$*, if* $V' \subseteq V$ *and* $E' \subseteq E$*. If G' contains all edges of G that join two vertices in V' then G' is said to be the graph* induced *by V'.*

**Definition 2.5** *Let* $W \subseteq V$*, then* $G - W$ *is the subgraph obtained by deleting the vertices in W from G, together with the edges incident to them.*

**Definition 2.6** *A* path *is a sequence of edges that connect a sequence of distinct vertices. A path between two vertices u and v is called a uv-path. If a uv-path exists between all pairs of vertices* $u,v \in V$*, then G is said to be* connected; *otherwise it is* disconnected.

**Definition 2.7** *The* length *of a uv-path* $P = (u = v_1, v_2, \ldots, v_l = v)$*, is the number of edges it contains, equal to* $l - 1$*. The* distance *between two vertices u and v is the minimal path length between u and v.*

**Definition 2.8** *A* cycle *is a uv-path for which* $u = v$*. All other vertices in the cycle must be distinct. A graph containing no cycles is said to be* acyclic.

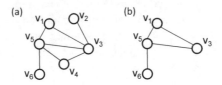

**Fig. 2.1** (a) A graph $G$, and (b) a subgraph $G'$ of $G$

To illustrate these definitions, Figure 2.1(a) shows a graph $G$ where, for example, vertices $v_1$ and $v_3$ are adjacent, but $v_1$ and $v_2$ are nonadjacent. The neighbourhood of $v_1$ is $\Gamma(v_1) = \{v_3, v_5\}$, giving $\deg(v_1) = 2$. The density of $G$ is $7/(1/2 \times 6 \times 5) = 0.467$. The subgraph $G'$ in Figure 2.1(b) has been created via the operation $G - \{v_2, v_4\}$, and in this case both $G$ and $G'$ are connected. Paths in $G$ from, for example, $v_1$ to $v_6$ include $(v_1, v_3, v_4, v_5, v_6)$ (of length 4) and $(v_1, v_5, v_6)$ (of length 2). Since the latter path is also the shortest path between $v_1$ to $v_6$, the distance between these vertices is also 2. Cycles also exist in both $G$ and $G'$, such as $(v_1, v_3, v_5, v_1)$.

## 2.1 The Greedy Algorithm

Recall the example from Section 1.1.1 where we sought to partition some students into a minimal number of groups for a team building exercise. The process we used to try and achieve this is known as the GREEDY algorithm, which is one of the simplest but most fundamental heuristic algorithms for graph colouring. The algorithm operates by taking vertices one by one according to some (possibly arbitrary) ordering and assigns each vertex its first available colour. Because this is a heuristic algorithm, the solutions it produces may very well be suboptimal; however, it can also be shown that GREEDY can produce an optimal solution for any graph given the correct sequence of vertices (see Theorem 2.2 below). As a result, various algorithms for graph colouring have been proposed that seek to find such orderings of the vertices (see Chapter 3).

---

GREEDY $(S \leftarrow \emptyset, \pi)$

---

(1) **for** $i \leftarrow 1$ **to** $|\pi|$ **do**
(2)         **for** $j \leftarrow 1$ **to** $|S|$
(3)                 **if** $(S_j \cup \{\pi_i\})$ is an independent set **then**
(4)                         $S_j \leftarrow S_j \cup \{\pi_i\}$
(5)                         **break**
(6)                 **else** $j \leftarrow j + 1$
(7)         **if** $j > |S|$ **then**
(8)                 $S_j \leftarrow \{\pi_i\}$
(9)                 $S \leftarrow S \cup S_j$

---

**Fig. 2.2** The GREEDY algorithm for graph colouring

Pseudocode for the GREEDY algorithm is given in Figure 2.2. To start, the algorithm takes an empty solution $S = \emptyset$ and an arbitrary permutation of the vertices $\pi$. In each outer loop the algorithm takes the $i$th vertex in the permutation, $\pi_i$, and attempts to find a colour class $S_j \in S$ into which it can be inserted. If such a colour class currently exists in $S$, then the vertex is added to it and the process moves on to consider the next vertex $\pi_{i+1}$. If not, lines (8–9) of the algorithm are used to create a new colour class for the vertex. An example run of the algorithm on a small graph is shown in Figure 2.3.

Let us now estimate the computational complexity of the GREEDY algorithm with regard to the number of constraint checks that are performed. We see that one vertex is coloured at each iteration, meaning $n = |\pi|$ iterations of the algorithm are required in total. At the $i$th iteration $(1 \leq i \leq n)$, we are concerned with finding a feasible colour for the vertex $\pi_i$. In the worst case this vertex will clash with all vertices that have preceded it in $\pi$, meaning that $(i - 1)$ constraint checks will be performed before a suitable colour is determined. Indeed, if the graph we are colouring is the complete graph $K_n$, the worst case will occur for all vertices; hence

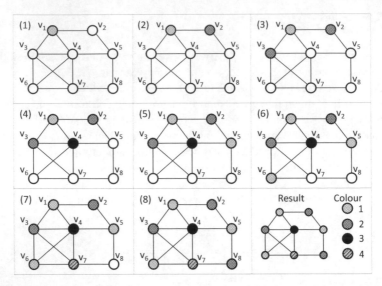

**Fig. 2.3** Example application of GREEDY using the permutation $\pi_i = v_i$ $(1 \leq i \leq n)$. Here, uncoloured vertices are shown in white

a total of $0 + 1 + 2 + \ldots + (n-1)$ constraint checks will be performed. This gives GREEDY an overall worst-case complexity $\mathcal{O}(n^2)$.

In practice, the GREEDY algorithm produces feasible solutions quite quickly; however, these solutions can often be quite poor in terms of the number of colours that the algorithm requires compared to the chromatic number. Consider, for example, the bipartite graph $G = (V_1, V_2, E)$ in which $n$ is even and where the vertex sets and edge set are defined $V_1 = \{v_1, v_3, \ldots, v_{n-1}\}$, $V_2 = \{v_2, v_4, \ldots, v_n\}$, and $E = \{\{v_i, v_j\} : v_i \in V_1 \wedge v_j \in V_2 \wedge i + 1 \neq j\}$. Figure 2.4 shows examples of such a graph using $n = 10$. Clearly for $n \geq 4$ such graphs will have a chromatic number $\chi(G) = 2$ because $V_1$ and $V_2$ constitute independent sets. However, colouring this graph using GREEDY with the permutation $\pi = (v_1, v_2, v_3, \ldots, v_n)$ will actually lead to a solution using $n/2$ colours, as Figure 2.4(a) illustrates. On the other hand, a permutation of the form $\pi = (v_1, v_3, \ldots, v_{n-1}, v_2, v_4, \ldots, v_n)$ will give the optimal solution shown in Figure 2.4(b). Clearly then, the order that the vertices are fed into the GREEDY algorithm can be very important.

One very useful feature of the GREEDY algorithm involves using existing feasible colourings of a graph to help generate new permutations of the vertices which can then be fed back into the algorithm. Consider the situation where we have a feasible colouring $\mathcal{S}$ of a graph $G$. Consider further a permutation $\pi$ of $G$'s vertices that has been generated such that the vertices occurring in each colour class of $\mathcal{S}$ are placed into adjacent locations in $\pi$. If we now use this permutation with GREEDY, the result will be a new solution $\mathcal{S}'$ that uses no more colours than $\mathcal{S}$, but possibly fewer. This is stated more concisely as follows:

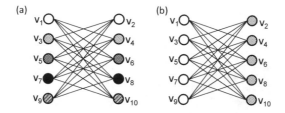

**Fig. 2.4** Two different colourings of a bipartite graph achieved by the GREEDY algorithm

**Theorem 2.1** *Let S be a feasible colouring of a graph G. If each colour class $S_i \in S$ (for $1 \le i \le |S|$) is considered in turn, and all vertices are fed one by one into the greedy algorithm, the resultant solution $S'$ will also be feasible, with $|S'| \le |S|$.*

*Proof.* Because $S = \{S_1, \ldots S_{|S|}\}$ is a feasible solution, each set $S_i \in S$ is an independent set. Obviously any subset $T \subseteq S_i$ is also an independent set. Now consider an application of GREEDY using $S$ to build a new candidate solution $S'$. In applying this algorithm, each set $S_1, \ldots, S_{|S|}$ is considered in turn, and all vertices $v \in S_i$ are assigned one by one to some set $S'_j \in S'$ according to the rules of GREEDY (that is, $v$ is first considered for inclusion in $S'_1$, then $S'_2$, and so on). Considering each vertex $v \in S_i$, two situations and resultant actions will occur in the following order of priority:

Case 1: An independent set $S'_{j<i} \in S'$ exists such that $S'_j \cup \{v\}$ is also an independent set. In this case $v$ will be assigned to the $j$th colour class in $S'$.

Case 2: An independent set $S'_{j=i} \in S'$ exists such that $S'_j \cup \{v\}$ is also an independent set.

In both cases it is clear that $v$ will always be assigned to a set in $S'$ with an index that is less than or equal to that of its original set in $S$. Of course, if a situation arises by which *all* items in a particular set $S_i$ are assigned according to Case 1, then at termination of GREEDY, $S'$ will contain fewer colours than $S$.

Now assume that it is necessary to assign a vertex $v \in S_i$ to a set $S'_{j>i}$. For this to occur, it is first necessary that the proposed actions of Cases 1 and 2 (i.e., adding $v$ to a set $S'_{j \le i}$) cause a clash. However, $S'_i \subset S_i$ and is therefore an independent set. By definition, $S'_i \cup \{v\} \subseteq S_i$ is also an independent set, contradicting the assumption. □

To show these concepts in action, the colouring shown in Figure 2.5(a) has been generated by the GREEDY algorithm using the permutation $\pi = (v_1, v_2, v_3, v_4, v_5, v_6, v_7, v_8)$, giving the 4-colouring $S = \{\{v_1, v_4, v_8\}, \{v_2, v_7\}, \{v_3, v_5\}, \{v_6\}\}$. This solution might then be used to form a new permutation $\pi = (v_1, v_4, v_8, v_2, v_7, v_3, v_5, v_6)$ which could then be fed back into the algorithm. However, our use of sets in defining a solution $S$ means that we are free to use any ordering of the colour classes in $S$ to form $\pi$, and indeed any ordering of the vertices within each colour class. One alternative permutation of the vertices formed from solution $S$ in this way is therefore $\pi = (v_2, v_7, v_5, v_3, v_6, v_4, v_8, v_1)$. This permutation has been used with GREEDY

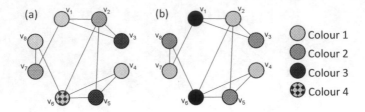

**Fig. 2.5** Feasible 4- and 3-colourings of a graph

to give the solution shown in Figure 2.5(b), which we see is using fewer colours than the solution from which it was formed.

These concepts give rise to the following theorem:

**Theorem 2.2** *Let $G$ be graph with an optimal graph colouring solution $S = \{S_1, \ldots, S_k\}$, where $k = \chi(G)$. Then there are at least*

$$\chi(G)! \prod_{i=1}^{\chi(G)} |S_i|! \tag{2.1}$$

*permutations of the vertices which, when fed into* GREEDY, *will result in an optimal solution.*

*Proof.* This arises immediately from Theorem 2.1: Since $S$ is optimal, an appropriate permutation can be generated from $S$ in the manner just described. Moreover, because the colour classes and vertices within each colour class can themselves be permuted, the above formula holds.                                                      □

Note that if $\chi(G) = 1$ or $\chi(G) = n$ then, trivially, the number of permutations decoding into an optimal solution will be $n!$. That is, every permutation of the vertices will decode to an optimal colouring using GREEDY.

## 2.2  Bounds on the Chromatic Number

In this section we now review some of the upper and lower bounds that can be stated about the chromatic number of a graph. Some of the bounds that we cover make use of the GREEDY algorithm in their proofs, helping us to further understand the behaviour of the algorithm. While these upper and lower bounds can be quite useful, or even exact for some topologies, we will see that in many cases they are either too difficult to calculate, or give us bounds that are too inaccurate to be of any practical use. This latter point will be demonstrated empirically later in Section 2.5.

### 2.2.1 Lower Bounds

To start, we make the observation that if a graph $G$ contains as a subgraph the complete graph $K_k$, then a feasible colouring of $G$ will obviously require at least $k$ colours. Stating this in another way, let $\omega(G)$ denote the number of vertices contained in the largest clique in $G$ (this is sometimes known as $G$'s *clique number*). Since $\omega(G)$ different colours will be needed to colour this clique, we deduce that $\chi(G) \geq \omega(G)$.

From another perspective, we can also consider the independent sets of a graph. Let $\alpha(G)$ denote the *independence number* of a graph $G$, defined as the number of vertices contained in the largest independent set in $G$. In this case, $\chi(G)$ must be at least $\lceil n/\alpha(G) \rceil$ since to be less than this value would imply the existence of an independent set larger than $\alpha(G)$.

These two bounds can be combined into the following:

$$\chi(G) \geq \max\{\omega(G), \lceil n/\alpha(G) \rceil\} \tag{2.2}$$

The accuracy of the bounds given in Inequality (2.2) will vary on a case to case basis. Their major drawback is the fact that the tasks of calculating $\omega(G)$ and $\alpha(G)$ are themselves NP-hard problems, namely the maximum clique problem and the maximum independent set problem. However, this does not mean that the bounds are useless: in some practical applications the sizes of the largest cliques and/or independent sets might be quite obvious from the graph's topology, or even specified as part of the problem itself (see, for example, the sport scheduling models used in Chapter 7). In other cases, we might also be able to approximate $\omega(G)$ and/or $\alpha(G)$ using heuristics or by applying probabilistic arguments.

To illustrate how we might estimate the size of a maximum clique in probabilistic terms, consider a graph $G$ with $n$ vertices that has been generated such that each pair of vertices is joined by an edge with probability $p$. Assuming independence, the probability that a subset of $x \leq n$ vertices forms a clique $K_x$ is calculated to be $p^{\binom{x}{2}}$, since there are $\binom{x}{2}$ edges that are required to be present among the $x$ vertices. The probability that the $x$ vertices do not form a clique is therefore simply $1 - p^{\binom{x}{2}}$. Since there are $\binom{n}{x}$ different subsets of $x$ vertices in $G$, the probability that none of these are cliques is calculated to be $(1 - p^{\binom{x}{2}})^{\binom{n}{x}}$. Hence the probability that there exists at least one clique of size $x$ in G is defined as

$$P(\exists K_x \subseteq G) = 1 - (1 - p^{\binom{x}{2}})^{\binom{n}{x}} \tag{2.3}$$

for $2 \leq x \leq n$. In practice we might use this formula to estimate a lower bound with a certain confidence. For example, we might say "with greater than 99% confidence we can say that $G$ contains a clique of size $y$", where $y$ represents the largest $x$ value for which Equation 2.3 is greater than 0.99. We might also collect similar information on the size of the largest maximum independent set in $G$ by simply replacing $p$ with $p = (1 - p)$ in the above formula. We must be careful in calculating the latter, however, because dividing $n$ by an underestimation of $\alpha(G)$ could lead to an invalid bound that exceeds $\chi(G)$. We should also be mindful that, for larger graphs, the numbers involved in calculating Equation (2.3) might be very large indeed, perhaps requiring rounding and introducing inaccuracies.

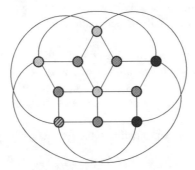

**Fig. 2.6** Optimal 4-colouring of the Grötzch graph

Even if we are able to estimate or determine values such as $\omega(G)$, we must still bear in mind that they may still constitute a very weak lower bound in many cases. Consider, for example, the graph shown in Figure 2.6, known as the Grötzch graph. This graph is considered "triangle free" in that it contains no cliques of size 3 or above; hence $\omega(G) = 2$. However, as illustrated in the figure, the chromatic number of the Grötzch graph is four: double the lower bound determined by $\omega(G)$. In fact, the Grötzch graph is the smallest graph in a set graphs known as the Mycielskians, named after their discoverer Jan Mycielski (1955). Mycielskian graphs demonstrate the potential inaccuracies involved in using $\omega(G)$ as a lower bound by showing that for any $q \geq 1$ there exists a graph $G$ with $\omega(G) = 2$ but for which $\chi(G) > q$. Hence we can encounter graphs for which $\omega(G)$ gives us a lower bound of 2, but for which the chromatic number can actually be arbitrarily large.

### 2.2.1.1 Bounds on Interval Graphs

While topologies such as the Mycielskian graphs demonstrate the potential for $\omega(G)$ to produce very poor lower bounds, in other cases this bound turns out to be both exact and easy to calculate. One practical application where this occurs is with *interval graphs*. Given a set of intervals defined on the real line, an interval graph is defined as a graph in which adjacent vertices correspond to overlapping intervals. More formally:

**Definition 2.9** *Let $\mathcal{I} = \{I_1, \ldots, I_n\}$ be a set of intervals defined on the real line such that each interval $I_i = \{x \in \mathbb{R} : a_i \leq x < b_i\}$, where $a_i$ and $b_i$ define the start and end values of interval $I_i$. The* interval graph of $\mathcal{I}$ *is the graph $G = (V, E)$ for which $V = \{v_1, \ldots, v_n\}$ and where $E = \{\{v_i, v_j\} : I_i \cap I_j \neq \emptyset\}$.*

An example interval graph has already been provided in Section 1.1.3 where we sought to assign taxi journeys with known start and end times to a minimal number of vehicles. Figure 1.5(a) in this section shows ten taxi journeys corresponding to ten intervals over the real line (representing time in this case). These intervals are then used to construct the interval graph shown in Figure 1.5(b).

One feature of interval graphs are that they are known to contain a "perfect elimination ordering". This is defined as an ordering of the vertices such that, for every vertex, all of its neighbours to the left of it in the ordering form a clique.

**Theorem 2.3** *Every interval graph G has a perfect elimination ordering.*

*Proof.* To start, arrange the intervals of $\mathcal{I}$ in ascending order of start values, such that $a_1 \leq a_2 \leq \ldots \leq a_n$. Now label the vertices $v_1, v_2, \ldots, v_n$ to correspond to this ordering. This implies that for any vertex $v_i$, the corresponding intervals of all neighbours to its left in the ordering must contain the value $a_i$; hence all pairs of $v_i$'s neighbours must also share an edge, thereby forming a clique.                                            □

The presence of a perfect elimination ordering is demonstrated in the example interval graph in Figure 1.5. Here we see, for example, that $v_3$ forms a clique of size 3 with its neighbours $v_2$ and $v_1$, and that $v_{10}$ forms a clique of size 2 with its neighbour $v_9$. The fact that all interval graphs contain a perfect elimination ordering allows us to produce optimal colourings to such graphs according to the following theorem.

**Theorem 2.4** *Let G be a graph with a perfect elimination ordering. An optimal colouring for G is obtained by labelling the vertices $v_1, \ldots, v_n$ such that $a_1 \leq a_2 \leq \ldots \leq a_n$, and then applying the GREEDY algorithm with the permutation $\pi_i = v_i$, for $i \leq i \leq n$. Moreover, $\chi(G) = \omega(G)$.*

*Proof.* During execution of GREEDY, each vertex $v_i = \pi_i$ is assigned to the lowest indexed colour not used by any of its neighbours preceding it in $\pi$. Clearly, each vertex has less than $\omega(G)$ neighbours. Hence at least one of the colours labelled 1 to $\omega(G)$ must be feasible for $v_i$. This implies $\chi(G) \leq \omega(G)$. Since $\omega(G) \leq \chi(G)$, this gives $\chi(G) = \omega(G)$.                                            □

The optimal 3-colouring provided in Figure 1.5(c) shows the result of this colouring process using the permutation $\pi = (v_1, v_2, v_3, v_4, v_5, v_6, v_7, v_8, v_9, v_{10})$.

More generally, graphs featuring perfect elimination orderings are usually known as *chordal* graphs. All interval graphs are therefore a type of chordal graph. The problem of determining whether a graph is chordal or not can be achieved in polynomial time by algorithms such as lexicographic breadth-first search (Rose et al., 1976). Hence any chordal graph can be recognised and coloured optimally in polynomial time.

## 2.2.2 Upper Bounds

Upper bounds on the chromatic number are often derived by considering the degrees of vertices in a graph. For instance, when a graph has a high density (that is, a high proportion of vertex pairs that are neighbouring), often a larger number of colours will be needed because a greater proportion of the vertex pairs will need to be separated into different colour classes. This admittedly rather weak-sounding proposal gives rise to the following theorem.

**Theorem 2.5** *Let $G$ be a connected graph with maximal degree $\Delta(G)$ (that is, $\Delta(G) = \max\{\deg(v) : v \in V\}$). Then $\chi(G) \leq \Delta(G) + 1$.*

*Proof.* Consider the behaviour of the GREEDY algorithm. Here the $i$th vertex in the permutation $\pi$, namely $\pi_i$, will be assigned to the lowest indexed colour class that contains none of its neighbouring vertices. Since each vertex has at most $\Delta(G)$ neighbours, no more than $\Delta(G) + 1$ colours will be needed to feasibly colour all vertices of $G$.                                                                                □

Another bound concerning vertex degrees can be calculated by examining all of a graph's subgraphs and identifying the minimal degree in each case, and then taking the maximum of these. For practical purposes this might be less useful than Theorem 2.5 for computing bounds quickly since the total number of subgraphs to consider might be prohibitively large. However, the following result still has its uses, particularly when it comes to colouring planar graphs and graphs representing circuit boards (see Chapter 5).

**Theorem 2.6** *Given a graph $G$, suppose that in every subgraph $G'$ of $G$ there is a vertex with degree less than or equal to $\delta$. Then $\chi(G) \leq \delta + 1$*

*Proof.* We know there is a vertex with a degree of at most $\delta$ in $G$. Call this vertex $v_n$. We also know that there is a vertex of at most $\delta$ in the subgraph $G - \{v_n\}$, which we can label $v_{n-1}$. Next, we can label as $v_{n-2}$ a vertex of degree of at most $\delta$ to form the graph $G - \{v_n, v_{n-1}\}$. Continue this process until all of the $n$ vertices have been assigned labels. Now assign these vertices to the permutation $\pi$ using $\pi_i = v_i$, and apply the GREEDY algorithm. At each step of the algorithm, $v_i$ will be adjacent to at most $\delta$ of the vertices $v_1, \ldots, v_{i-1}$ that have already been coloured; hence no more than $\delta + 1$ colours will be required.                                                □

Let us now examine some implications of these two theorems. It can be seen that Theorem 2.5 provides tight bounds for both complete graphs, where $\chi(K_n) = \Delta(K_n) + 1 = n$, and for odd cycles, where $\chi(C_n) = \Delta(C_n) + 1 = 3$. However, such accurate bounds will not always be so forthcoming. Consider, for example, the wheel graph comprising 100 vertices, $W_{100}$. This features a "central" vertex of degree $\Delta(W_{100}) = 99$, meaning that Theorem 2.5 merely informs us that the chromatic number of $W_{100}$ is less than 100, despite the fact that it is actually just four! On the other hand, for any wheel graph it is relatively easy to show that all of its subgraphs will contain a vertex with a degree of no more than 3 (i.e., $\delta = 3$). For wheel graphs where $n$ is even, this allows Theorem 2.6 to return a tight bound since $\delta + 1 = \chi(W_n) = 4$.

Beyond complete graphs and odd cycles, Theorem 2.5 can also be marginally strengthened due to the result of Brooks (1941). This proof is slightly more involved than that of Theorem 2.5 and requires some further definitions.

**Definition 2.10** *A* component *of a graph G is a subgraph $G'$ in which all pairs of vertices are connected by paths. A graph that is itself connected has exactly one component, comprising the whole graph.*

**Definition 2.11** *A* cut vertex *$v$ is a vertex whose removal from a graph G (together with all incident edges) increases the number of components. Thus a cut vertex of a connected graph is a vertex whose removal disconnects the graph. More generally, a* separating set *of a graph G is a set of vertices whose removal increases the number of components.*

**Definition 2.12** *A graph G is said to be $k$-connected if it remains connected whenever fewer than $k$ vertices are removed. In other words, G will only become disconnected if a* separating set *comprising $k$ or more vertices is deleted.*

**Definition 2.13** *A component of a graph is considered a* block *if it is 2-connected.*

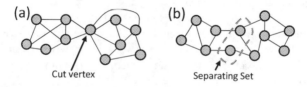

Cut vertex                    Separating Set

**Fig. 2.7** Illustrations of a cut vertex and separating set

To illustrate these definitions, Figure 2.7(a) shows a graph $G$ comprising one component. Removal of the indicated cut vertex would split $G$ into two components. Figure 2.7(b) can be considered a block in that it does not contain a cut vertex (i.e., it is 2-connected). However, it is not 3-connected, because removal of the two vertices in the indicated separating set increases the number of components from one to two.

Having gone over the necessary terminology, we are now in a position to state and prove Brooks' theorem.

**Theorem 2.7 (Brooks (1941))** *Let G be a connected graph with maximal degree $\Delta(G)$. Suppose further that G is not complete and not an odd cycle. Then $\chi(G) \leq \Delta(G)$.*

*Proof.* The theorem is obviously correct for $\Delta(G) \leq 2$. For $\Delta(G) = 0$ and $\Delta(G) = 1$, the corresponding graphs will be $K_1$ and $K_2$ respectively, and are therefore not included in the theorem. For $\Delta(G) = 2$ on the other hand, $G$ will be a path or even cycle (giving $\chi(G) = 2$) or will be an odd cycle, meaning it is not included in the theorem.

Assuming $\Delta(G) \geq 3$, let $G$ be a counterexample with the smallest possible number of vertices for which the theorem does not hold: i.e., $\chi(G) > \Delta(G)$. We therefore assume that all graphs with fewer vertices than $G$ can be feasibly coloured using $\Delta(G)$ colours.

Claim 1:   $G$ is connected. If $G$ were not connected, then $G$'s components would be a smaller counterexample, or all of $G$'s components would be $\Delta(G)$-colourable.

Claim 2:   $G$ is 2-connected. If $G$ were not 2-connected, then $G$ would have at least one cut vertex $v$, and each block of $G$ would be $\Delta(G)$-colourable. The colourings of each block could then be combined to form a feasible $\Delta(G)$-colouring.

Claim 3:   $G$ must contain three vertices $v$, $u_1$ and $u_2$ such that (a) $u_1$ and $u_2$ are non-adjacent; (b) both $u_1$ and $u_2$ are adjacent to $v$; and (c) $G - \{u_1, u_2\}$ is connected. Two cases can now be considered:

Case 1:   $G$ is 3-connected. Because $G$ is not complete, there must be two vertices $x$ and $y$ that are nonadjacent. Let the shortest path between $x$ and $y$ in $G$ be $x = v_0, \ldots, v_l = y$, where $l \geq 2$. Since this is the shortest path, $v_0$ is not adjacent to $v_2$, so we can choose $u_1 = v_0$, $v = v_1$ and $u_2 = v_2$. This satisfies Claim 3.

Case 2:   $G$ is 2-connected but not 3-connected. In this case, there must exist two vertices $u$ and $v$ such that $G - \{u, v\}$ is disconnected. This means that the graph $G - \{v\}$ contains a cut vertex (i.e., $u$), but there is no cut vertex in $G$ itself. In this case, $v$ must be adjacent to at least one vertex in every block of the graph $G - \{v\}$. Let $u_1$ and $u_2$ be two vertices in two different end blocks of $G - \{v\}$ that are adjacent to $v$. The vertices $u_1$, $u_2$ and $v$ now satisfy Claim 3.

Having proved Claims 1, 2, and 3, we now construct a permutation $\pi$ of the $n$ vertices of $G$ such that $\pi_1 = u_1$, $\pi_2 = u_2$, and $\pi_n = v$. The remaining parts of the permutation $\pi_3, \ldots \pi_{n-1}$ are then formed such that, for $3 \leq i < j \leq n-1$, the distance from $\pi_n$ to $\pi_i$ is greater than or equal to the distance from $\pi_n$ to $\pi_j$. If we now apply GREEDY to this permutation, the vertices $\pi_1 = u_1$ and $\pi_2 = u_2$ will first both be assigned to colour $S_1$, because they are nonadjacent. Moreover, when we colour the vertices $\pi_i$ ($3 \leq i < n$), there will always be at least one colour $S_{j \leq \Delta(G)}$ available for $\pi_i$. Finally, when we come to colour vertex $\pi_n = v$, at most $\Delta(G) - 1$ colours will have been used to colour the neighbours of $v$ (since its neighbours $u_1$ and $u_2$ have been assigned to the same colour) and so at least one of the $\Delta(G)$ colours will be feasible for $v$. This shows that $\chi(G) \leq \Delta(G)$ as required.                    □

Having analysed the behaviour of the GREEDY algorithm and reviewed a number of bounds on the chromatic number, the following two sections will now consider two further heuristic-based constructive algorithms for the graph colouring problem. As we will see, these algorithms are guaranteed to produce optimal solutions for some simple graph topologies and also often construct solutions that improve on the upper bounds mentioned above. Later, in Chapters 3 and 4, we will also see that these algorithms, along with GREEDY, are often used as building blocks in many of the more sophisticated algorithms available for graph colouring.

## 2.3 The DSATUR Algorithm

The DSATUR algorithm (abbreviated from "degree of saturation") was originally proposed by Brélaz (1979). In essence it is very similar in behaviour to the GREEDY algorithm in that it takes each vertex in turn according to some ordering and then assigns it to the first suitable colour class, creating new colour classes when necessary. The difference between the two algorithms lies in the way that these vertex orderings are generated. With GREEDY the ordering is decided before any colouring takes place; on the other hand, for the DSATUR algorithm the choice of which vertex to colour next is decided heuristically based on the characteristics of the current partial colouring of the graph. This choice is based primarily on the *saturation degree* of the vertices, defined as follows.

**Definition 2.14** *Let* $c(v) =$ NULL *for any vertex* $v \in V$ *not currently assigned to a colour class. Given such a vertex* $v$, *the* saturation degree *of* $v$, *denoted by* $\mathrm{sat}(v)$, *is the number of different colours assigned to adjacent vertices. That is,* $\mathrm{sat}(v) = |\{c(u) : u \in \Gamma(v) \wedge c(u) \neq$ NULL$\}|$

| DSATUR $(S \leftarrow \emptyset, X \leftarrow V)$ |
| --- |
| (1) **while** $X \neq \emptyset$ **do** |
| (2)     Choose $v \in X$ |
| (3)     **for** $j \leftarrow 1$ **to** $|S|$ |
| (4)         **if** $(S_j \cup \{v\})$ is an independent set **then** |
| (5)             $S_j \leftarrow S_j \cup \{v\}$ |
| (6)             **break** |
| (7)         **else** $j \leftarrow j + 1$ |
| (8)     **if** $j > |S|$ **then** |
| (9)         $S_j \leftarrow \{v\}$ |
| (10)        $S \leftarrow S \cup S_j$ |
| (11)    $X \leftarrow X - \{v\}$ |

**Fig. 2.8** The DSATUR algorithm for graph colouring

Pseudocode for the DSATUR algorithm is shown in Figure 2.8. It can be seen that the majority of the algorithm is the same as the GREEDY algorithm in that once a vertex has been selected, a colour is found by simply going through each colour class in turn and stopping when a suitable one has been found. Consequently, the worst-case complexity of DSATUR is the same as GREEDY at $\mathcal{O}(n^2)$, although in practice some extra bookkeeping is required to keep track of the saturation degrees of the uncoloured vertices.

The major difference between GREEDY and DSATUR lies in lines (1), (2) and (11) of the pseudocode. Here, a set $X$ is used to define the set of vertices currently not assigned to a colour. At the beginning of execution $X = V$. In each iteration of the algorithm the next vertex to be coloured is selected from $X$ according to line (2), and once coloured, it is removed from $X$ in line (11). The algorithm terminates when $X = \emptyset$.

Line (2) of Figure 2.8 provides the main power behind the DSATUR algorithm. Here, the next vertex to be coloured is chosen as the vertex in $X$ that has maximal saturation degree. If there is more than one vertex with maximal saturation degree, then ties are broken by choosing the vertex among these with the largest degree. Any further ties can then be broken randomly. The idea behind the maximum saturation degree heuristic is that it prioritises vertices that are seen to be the most "constrained"—that is, vertices that currently have the fewest colour options available to them. Consequently, these "more constrained" vertices are dealt with by the algorithm first, allowing the less constrained vertices to be coloured later.

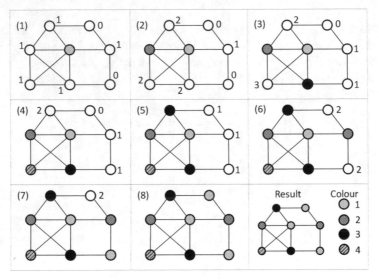

**Fig. 2.9** Example application of DSATUR. Here, uncoloured vertices (members of $X$) are shown in white, and have their saturation degrees written alongside

Figure 2.9 shows an example run-through of the algorithm on a small graph. To begin, all vertices have a saturation degree equal to 0, so the first vertex to be coloured is the one with the highest degree. As shown in Step (1), this is assigned to colour 1. This also leads to five vertices having a saturation degree of 1, so the next vertex to be chosen is the one among these that has the highest degree. This is then assigned to colour 2 as shown in Step (2). Next, three vertices have saturation degrees of 2, so we again choose the vertex among these with the highest degree. Since colours 1 and 2 are not feasible for this vertex, it is assigned to colour 3.

This process continues as shown in the figure until a feasible colouring has been achieved.

Earlier we saw that the number of colours used in solutions produced by the GREEDY algorithm depends on the order that the vertices are fed into the procedure, with results (in terms of the number of colours used in the solution produced) potentially varying a great deal. On the other hand, the DSATUR algorithm reduces this variance by generating the vertex ordering *during* a run according to its heuristics. As a result, DSATUR's performance is more predictable. Indeed DSATUR turns out to be exact for a number of elementary graph topologies. The first of these is the bipartite graph, and to prove this claim it is first necessary to show a classical result on the structure of these graphs.

**Theorem 2.8** *A graph is bipartite if and only if it contains no odd cycles.*

*Proof.* Let $G$ be a connected bipartite graph with vertex sets $V_1$ and $V_2$. (It is enough to consider $G$ as being connected, as otherwise we could simply treat each component of $G$ separately.) Let $v_1, v_2, \ldots, v_l, v_1$ be a cycle in $G$. We can also assume that $v_1 \in V_1$, $v_2 \in V_2$, $v_3 \in V_1$, and so on. Hence, a vertex $v_i \in V_1$ if and only if $i$ is odd. Since $v_l \in V_2$, this implies $l$ is even. Consequently $G$ has no odd cycles.

Now suppose that $G$ is known to feature no odd cycles. Choose any vertex $v$ in the graph and let the set $V_1$ be the set of vertices such that the shortest path from each member of $V_1$ to $v$ is of odd length, and let $V_2$ be the set of vertices where the shortest path from each member of $V_2$ to $v$ is even. Observe now that there is no edge joining vertices of the same set $V_i$ since otherwise $G$ would contain an odd cycle. Hence $G$ is bipartite. □

This result allows us to prove the following theorem.

**Theorem 2.9 (Brélaz (1979))** *The DSATUR algorithm is exact for bipartite graphs.*

*Proof.* Let $G$ be a connected bipartite graph with $n \geq 3$. If $G$ is not connected, it is enough to consider each component of $G$ separately. For purposes of contradiction assume that one vertex $v$ has a saturation degree of 2, meaning that $v$ has two neighbours, $u_1$ and $u_2$, assigned to different colours. From these two neighbours we can build two paths which, because $G$ is connected, will have a common vertex $u$. Hence we have formed a cycle containing vertices $v$, $u_1$, $u_2$, $u$ and perhaps others. Since $G$ is bipartite, the length of this cycle must be even, meaning that the $u_1$ and $u_2$ must have the same colour, contradicting our initial assumption. □

To illustrate the usefulness of this result, consider the bipartite graphs shown in Figure 2.4 earlier. Here, many permutations of the vertices used in conjunction with the GREEDY algorithm will lead to colourings using more than two colours. Indeed, in the worst case they may even lead to $(n/2)$-colourings as demonstrated in the figure. In contrast DSATUR is guaranteed to return the optimal solution bipartite graphs, as it is for some further topologies:

**Theorem 2.10** *The DSATUR algorithm is exact for cycle and wheel graphs.*

*Proof.* Note that even cycles are 2-colourable and are therefore bipartite. Hence they are dealt with by Theorem 2.9. However, it is useful to consider both even and odd cycles in the following.

Let $C_n$ be a cycle graph. Since the degree of all vertices in $C_n$ is 2, the first vertex to be coloured, $v$, will be chosen arbitrarily by DSATUR. In the next $(n-2)$ steps, according to the behaviour of DSATUR a path of vertices of alternating colours will be constructed that extends from $v$ in both clockwise and anticlockwise directions. At the end of this process, a path comprising $n-1$ vertices will have been formed, and a single vertex $u$ will remain that is adjacent to both terminal vertices of this path. If $C_n$ is an even cycle, $n-1$ will be odd, meaning that the terminal vertices have the same colour. Hence $u$ can be coloured with the alternative colour. If $C_n$ is an odd cycle, $n-1$ will be even, meaning that the terminal vertices will have different colours. Hence a $u$ will be assigned to a third colour.

For wheel graphs $W_n$ a similar argument applies. Assuming $n \geq 5$, DSATUR will initially colour the central vertex $v_n$ because it features the highest degree. Since $v_n$ is adjacent to all other vertices in $W_n$, all remaining vertices $v_1, \ldots, v_{n-1}$ will now have a saturation degree of 1. The same colouring process as the cycle graphs $C_{n-1}$ then follows.                                                                                           $\square$

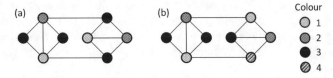

**Fig. 2.10** An optimal 3-colouring (a) and a suboptimal 4-colouring produced by DSATUR (b)

Although, as these theorems show, DSATUR is exact for certain types of graph, the NP-hardness of the graph colouring problem obviously implies that it will be unable to produce optimal solutions for *all* graphs. Figure 2.10, for example, shows a small graph that, while actually being 3-colourable, will always be coloured using four colours by DSATUR, regardless of the way any random ties in the algorithm's heuristics are broken. In fact, Janczewski et al. (2001) have proved that this is the smallest such graph where this suboptimality occurs, but there are countless larger graphs where DSATUR will also not return the optimal. In other work, Spinrad and Vijayan (1984) have also identified a graph topology of $\mathcal{O}(n)$ vertices that, despite being 3-colourable, will actually be coloured using $n$ different colours using DSATUR.

## 2.4 The Recursive Largest First (RLF) Algorithm

While the DSATUR algorithm for graph colouring is similar in behaviour and complexity to the classical GREEDY approach, the next constructive method we examine, the Recursive Largest First (RLF) algorithm, follows a slightly different strategy. The RLF algorithm was originally designed by Leighton (1979), in part for use in constructing solutions to large timetabling problems. The method works by colouring a graph *one colour at a time*, as opposed to one vertex at a time. In each step the algorithm uses heuristics to identify an independent set of vertices in the graph, which are then associated with the same colour. This independent set is then removed from the graph, and the process is repeated on the resultant, smaller subgraph. This process continues until the subgraph is empty, at which point all vertices have been coloured leaving us with a feasible solution. Leighton (1979) has proven the worst-case complexity of RLF to be $\mathcal{O}(n^3)$, giving it a higher computational cost than the $\mathcal{O}(n^2)$ GREEDY and DSATUR algorithms; however, this algorithm is still of course polynomially bounded.

| RLF $(S \leftarrow \emptyset, X \leftarrow V, Y \leftarrow \emptyset, i \leftarrow 0)$ |
|---|
| (1) **while** $X \neq \emptyset$ **do** |
| (2) $\quad i \leftarrow i+1$ |
| (3) $\quad S_i \leftarrow \emptyset$ |
| (4) $\quad$ **while** $X \neq \emptyset$ **do** |
| (5) $\quad\quad$ Choose $v \in X$ |
| (6) $\quad\quad S_i \leftarrow S_i \cup \{v\}$ |
| (7) $\quad\quad Y \leftarrow Y \cup \Gamma_X(v)$ |
| (8) $\quad\quad X \leftarrow X - (Y \cup \{v\})$ |
| (9) $\quad S \leftarrow S \cup \{S_i\}$ |
| (10) $\quad X \leftarrow Y$ |
| (11) $\quad Y \leftarrow \emptyset$ |

**Fig. 2.11** The RLF algorithm for graph colouring. Here, $\Gamma_X(v)$ denotes the subset of vertices in the set $X$ that are adjacent to vertex $v$

Pseudocode for the RLF algorithm is given in Figure 2.11. In each outer loop of the process, the $i$th colour class $S_i$ is build. The algorithm also makes use of two sets: $X$, which contains uncoloured vertices that can currently be added to $S_i$ without causing a clash; and $Y$, which holds the uncoloured vertices that *cannot* be feasibly added to $S_i$. At the start of execution $X = V$ and $Y = \emptyset$.

Lines (4) to (8) of Figure 2.11 give the steps responsible for constructing the $i$th colour class $S_i$. To start, a vertex $v$ from $X$ is selected and added to $S_i$ (i.e., $v$ is coloured with colour $i$). Next, all vertices neighbouring $v$ in the subgraph induced by $X$ are transferred to $Y$, to signify that they cannot now be feasibly assigned to $S_i$. Finally, $v$ and its neighbours are also removed from $X$, since they are not now considered candidates for inclusion in colour class $S_i$.

Once $X = \emptyset$, no further vertices can be added to the current colour class $S_i$. In lines (9) to (11) of the algorithm $S_i$ is therefore added to the solution $S$ and, if necessary, the algorithm moves on to constructing colour class $S_{i+1}$. To do this, all vertices in the set of uncoloured vertices $Y$ are moved into $X$, and $Y$ is emptied. Obviously, once both $X$ and $Y$ are empty, all vertices have been coloured.

The heuristics suggested by Leighton (1979) for selecting the next vertex $v \in X$ to colour in line (5) follow a similar rationale to those of the DSATUR algorithm in that the most "constrained" vertices are prioritised. Consequently the first vertex chosen for insertion into each colour class $S_i$ is the member of $X$ that has the highest degree in the subgraph induced by $X$. The remaining vertices $v$ for $S_i$ are then selected as the member of $X$ that has the largest degree in the subgraph induced by $Y \cup \{v\}$ (that is, the vertex in $X$ that is adjacent to the largest number of vertices in $Y$). As with DSATUR, any ties in these heuristics can be broken randomly.

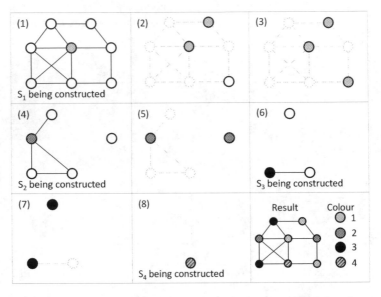

**Fig. 2.12** Example application of the RLF algorithm. Here, dotted vertices denote those currently assigned to the set $Y$. Vertices with solid lines and no colour denote members of $X$

Figure 2.12 gives an example step-by-step run-through of the RLF algorithm. Steps that involve the creation of a new colour class $S_i$ are indicated. In Step (1) the vertex $v$ with the highest degree in the graph is added to colour class $S_1$. In Step (2), all vertices adjacent to $v$ have now been moved to $Y$, leaving a subgraph induced by the set $X$ which contains just two vertices, both of which are subsequently added to colour class $S_1$ in Steps (2) and (3). In Step (4) a new colour class is created and the process is repeated on the subgraph induced by the remaining uncoloured vertices. This continues until all vertices have been coloured, as shown.

Like DSATUR, the RLF algorithm is also exact for a number of fundamental graph topologies.

**Theorem 2.11** *The* RLF *algorithm is exact for bipartite graphs.*

*Proof.* Let $G$ be a connected bipartite graph with $n \geq 3$ and vertex sets $V_1$ and $V_2$. If $G$ is disconnected, it is enough to consider each component separately.

Assume without loss of generality that vertex $v \in V_1$ has the highest degree and is therefore assigned to the first colour. Consequently all neighbours of $v$ (that is, $\Gamma(v) \subseteq V_2$) will be added to $Y$. It is now sufficient to show that the next $|V_1| - 1$ vertices assigned to the first colour will all be members of $V_1$. This is indeed the case because, in the next $|V_1| - 1$ steps, uncoloured vertices will be selected that have the largest number of adjacent vertices in the set $Y$. Since uncoloured vertices from the set $V_2 - Y$ will be nonadjacent to those in $Y$, only vertices from $V_1$ will be selected. This applies until all vertices from $V_1$ have been assigned to the first colour. At this point the subgraph induced by $V_2$ will have no edges, allowing RLF to colour all remaining vertices with the second colour. □

**Theorem 2.12** *The* RLF *algorithm is exact for cycle and wheel graphs.*

*Proof.* Even cycles are 2-colourable and are thus dealt with by Theorem 2.11. However, for convenience we shall consider both even and odd cycles in the following. Let $C_n$ be a cycle graph with vertices $V = \{v_1, \ldots, v_n\}$ and edges $E = \{\{v_1, v_2\}, \{v_2, v_3\}, \ldots, \{v_{n-1}, v_n\}, \{v_n, v_1\}\}$. For bookkeeping purposes, also assume that ties in the RLF selection heuristic (line (4) of Figure 2.11) are broken by taking the vertex with the lowest index, as opposed to choosing arbitrarily. It is easy to see that this theorem holds without this restriction, however.

The degree of all vertices in $C_n$ is 2, so the first vertex to be coloured will be $v_1$. Consequently, neighbouring vertices $v_2$ and $v_{n-1}$ are added to $Y$. According to the heuristics of RLF the next vertex to be coloured will be $v_3$, leading to $v_4$ being added to $Y$; then $v_5$, leading to $v_6$ being added to $Y$; and so on. At the end of this process, we will have colour class $S_1 = \{v_1, v_3, \ldots, v_{n-1}\}$ when $n$ is even, and the colour class $S_1 = \{v_1, v_3, \ldots, v_{n-2}\}$ when $n$ is odd. In the even case, this leaves an uncoloured subgraph with vertices $\{v_2, v_4, \ldots, v_n\}$ and no edges. Consequently RLF will assign all of these vertices to the second colour. In the odd case, we will be left with uncoloured vertices $\{v_2, v_4, \ldots, v_{n-1}, v_n\}$ together with a single edge $\{v_{n-1}, v_n\}$. Following the heuristic rules of RLF, all even-indexed vertices will then be assigned to the second colour, with $v_n$ being assigned to the third.

For wheel graphs $W_n$ similar reasoning applies. Assuming $n \geq 5$, the central vertex $v_n$ will coloured first because it has the highest degree. Since $v_n$ is adjacent to all other vertices, no further vertices can be added to this colour, so the algorithm will move on to the second colour. The remaining uncoloured vertices now form the cycle graph $C_{n-1}$, and the same colouring process as above follows. □

## 2.5 Empirical Comparison

In this section we now present a comparison of the GREEDY, DSATUR, and RLF algorithms looking particularly at their run time requirements and the quality of solutions that they tend to produce. The algorithm implementations used in these experiments can be found in the online suite of graph colouring algorithms mentioned in Section 1.6.1 and Appendix A.1.

## 2.5.1 Experimental Considerations

When new algorithms are proposed for the graph colouring problem, the quality of the solutions it produces will usually be compared to those achieved on the same problem instances by other preexisting methods. A development in this regard occurred in 1992 with the organisation of the Second DIMACS Implementation Challenge (http://mat.gsia.cmu.edu/COLOR/instances.html), which resulted in a suite of differently structured graph colouring problems being placed into the public domain. Since this time, authors of graph colouring papers have often used this set (or a subset of it) and have concentrated on tuning their algorithms (perhaps by altering the underlying heuristics or run time parameters) in order to achieve the best possible results.

More generally, when testing and comparing the accuracy of two graph colouring algorithms (or, indeed, any approximation/heuristic algorithms), an important question is "Are we attempting to show that Algorithm A produces better results than Algorithm B on (a) a particular problem instance, or (b) across a whole set of problem instances?" In some cases we might, for example, be given a difficult practical problem that we only need to solve once, and whose efficient solution might save lots of money or other resources. Here, it would seem sensible to concentrate on answering question (a) and spend our time choosing the correct heuristics and parameters in order to achieve the best solution possible under the given time constraints. If our chosen algorithm involves stochastic behaviour (i.e., making random choices) multiple runs of the algorithm might then be performed on the problem instance to gain an understanding of its average performance for this case. In most situations however, it is more likely that when a new algorithm is proposed, the scientific community will be more interested in question (b) being answered—that is, we will want to understand and appreciate the performance of the algorithm across a whole set of problem instances, allowing more general conclusions to be drawn.

If we choose to follow (b) above, it is first necessary to decide what *types* of graphs (i.e., what population of problems instances) we wish to make statements about. For instance, this might be the set of all 2-colourable graphs, or it could be the set of all graphs containing fewer than 1,000 vertices. Typically, populations like these will be very large, or perhaps infinite in size, and so it will be necessary to test our algorithms on randomly selected samples of these populations. Under appropriate experimental conditions, we might then be able to use the outcomes of these trials to make general statistical statements about the population itself, such as: "With $\geq 95\%$ confidence Algorithm A produces solutions with fewer colours than Algorithm B on this particular graph type".

In this section, in order to compare the performance of the GREEDY, DSATUR, and RLF algorithms, we make use of the following facts to define our population. Given a graph with $n$ vertices, let $l$ denote the number of vertex pairs in $G$. That is, $l = \binom{n}{2}$. Any graph with $n$ vertices can therefore be represented by an $l$-dimensional binary vector $\mathbf{b}^{(n)}$ for which element $b_i^{(n)} = 1$ if and only if the corresponding pair of vertices are adjacent, and $b_i^{(n)} = 0$ otherwise.

Now let $\mathcal{B}^{(n)}$ define the set of all $l$-dimensional binary vectors. Obviously this means that $|\mathcal{B}^{(n)}| = 2^l$. The set $\mathcal{B}^{(n)}$ can therefore be considered as the set of all possible ways of connecting the vertices in an $n$-vertex graph. However, we must be careful in this interpretation as this is not quite the same as saying that $\mathcal{B}^{(n)}$ represents the set of all *graphs* with $n$ vertices (which it does not), because it fails to take into account the principle of graph isomorphisms.

Consider the example in Figure 2.13, where we show two different six-dimensional binary vectors and illustrate the graphs that they represent, called $G_1$ and $G_2$ here. Note that when we come to colour $G_1$ and $G_2$ the vertex labels are of little importance and, indeed, without the labels the two graphs might be considered identical. In these circumstances $G_1$ and $G_2$ are considered *isomorphic* as there exists a way of converting one graph into the other by simply relabelling the vertices (in this example we can convert $G_1$ to $G_2$ by relabelling $v_1$ as $v_2$, $v_2$ as $v_4$, $v_3$ as $v_3$ and $v_4$ as $v_1$). Because the set $\mathcal{B}^{(n)}$ fails to take these isomorphisms into account, it must therefore be interpreted as the "set of all $n$-vertex graphs *and their isomorphisms*", as opposed to the set of all $n$-vertex graphs itself.

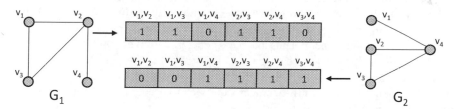

**Fig. 2.13** Illustration of how different binary vectors can represent graphs that are isomorphic

To generate a single member of the set $\mathcal{B}^{(n)}$ at random (i.e., to choose an element of $\mathcal{B}^{(n)}$ such that each element is equally likely to be selected), it is simply necessary to generate an $l$-dimensional vector $\mathbf{b}^{(n)}$ in which each element $b_i^{(n)} = 1$ with probability 0.5, and 0 otherwise. This is the same process as producing a *random graph* with $p = 0.5$:

**Definition 2.15** *A random graph, denoted by $G_{n,p}$, is a graph comprising $n$ vertices in which each pair of vertices is adjacent with probability $p$. The degrees of the vertices in a random graph are consequently binomially distributed:* $\deg(v) \sim B(n-1, p)$.

Random graphs will be the focus of our algorithm comparison in this chapter, though we will also look at other types of graphs in later chapters.

**Table 2.1** Summary of results produced by the GREEDY, DSATUR and RLF algorithms on random graphs $G_{n,0.5}$

| | | Algorithm[a] | | | |
|---|---|---|---|---|---|
| $n$ | LB[b] | GREEDY | DSATUR | RLF | UB[c] |
| 100 | 9 | $21.14 \pm 0.95$ | $18.48 \pm 0.81$ | $17.44 \pm 0.61$ | 62.22 |
| 500 | 10 | $72.54 \pm 1.33$ | $65.18 \pm 1.06$ | $61.04 \pm 0.78$ | 284.06 |
| 1000 | 10 | $126.64 \pm 1.21$ | $115.44 \pm 1.23$ | $108.74 \pm 0.90$ | 550.76 |
| 1500 | 10 | $176.20 \pm 1.58$ | $162.46 \pm 1.42$ | $153.44 \pm 0.86$ | 841.92 |
| 2000 | 10 | $224.18 \pm 1.90$ | $208.18 \pm 1.02$ | $196.88 \pm 1.10$ | 1076.26 |

[a] Mean plus/minus standard deviation in number of colours, taken from runs across 50 graphs.

[b] Largest value $x$ for which Equation (2.3) is greater than or equal to 0.99.

[c] Generated according to Theorem 2.7. Mean taken across 50 graphs.

## 2.5.2 Results and Analysis

Table 2.1 summarises the number of colours used in solutions produced by the GREEDY, DSATUR, and RLF algorithms for random graphs with edge probability $p = 0.5$ and varying numbers of vertices. For each value of $n$, 50 random graphs were generated and each algorithm was executed on it once. In applications of the GREEDY algorithm, the vertex permutation $\pi$ was generated randomly. Table 2.1 also shows lower and upper bounds that were generated for these problem instances.

The results in Table 2.1 indicate that for all tested values of $n$, the DSATUR algorithm tends to produce solutions using fewer colours than the GREEDY algorithm. Indeed, in all five cases, these differences were seen to be significantly different.[1] In turn, significant differences were also observed between the results of DSATUR and RLF, indicating that, for all of the tested values of $n$, the RLF algorithm produces the best solutions across the set of all graphs and their isomorphisms.

The data in Table 2.1 also reveals that the generated lower and upper bounds seem to be some distance from the number of colours ultimately used by the algorithms. This indicates that Brooks' Theorem (2.7) tends to provide a rather inaccurate upper bound for random graphs. It also suggests two factors with regard to the lower bound: (a) that the probabilistic bound determined by Equation (2.3) is also quite inaccurate and/or (b) that the GREEDY, DSATUR, and RLF algorithms are producing solutions whose numbers of colours are some distance from the chromatic number.

The graphs shown in Figure 2.3 expand upon the results of Table 2.1 by considering a range of different values for $p$. Bounds are also indicated by the shaded areas. We see that the unshaded areas of these graphs are generally quite wide, with the algorithms' results falling in a fairly narrow band within these. This again indicates the inadequacy of the upper bound, particularly for larger values of $n$.

The differences between the three algorithms themselves across these values of $p$ are presented more clearly in Figure 2.15. Here, the bars in the graphs show the

---

[1] The samples collected for each algorithm and value of $n$ were not generally found to be derived from an underlying normal distribution according to a Shapiro-Wilk test. Consequently, statistical significance is claimed here according to the results of a nonparametric related samples Wilcoxon Signed Rank test at the 0.1% significance level.

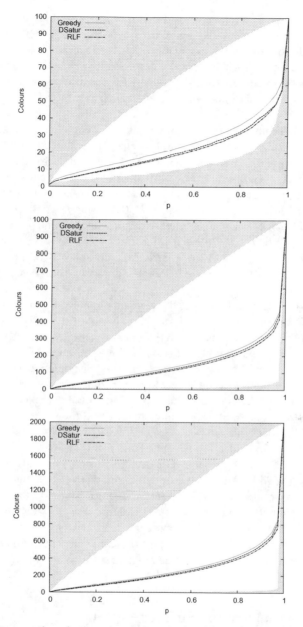

**Fig. 2.14** Average number of colours in solutions produced by the GREEDY, DSATUR and RLF algorithms on random graphs $G_{n,p}$ with various values of $p$, using $n = 100$, 1,000 and 2,000. All points are the mean across 50 graphs

number of colours used in solutions produced by the GREEDY algorithm, and the lines indicate the percentage of this figure used by DSATUR and RLF. We see that the latter two algorithms achieve percentages of less than 100% across all of the tested values for $p$, indicating their superior performance across the range of random graphs, from sparse to dense. We also see that RLF consistently produces the lowest percentages, once again indicating its general superiority to DSATUR on these graphs.

We now turn to the implications of the computational complexity of the GREEDY, DSATUR, and RLF algorithms. Earlier in this chapter we noted that GREEDY and DSATUR both have worst-case complexities of $\mathcal{O}(n^2)$, while for the RLF this is $\mathcal{O}(n^3)$. What effects does this have when running the algorithms on different graph colouring problems? Figure 2.16 shows the number of constraint checks required by the algorithms for random graphs with $p = 0.5$ using values of $n$ up to 2,000. For larger graphs the RLF algorithm clearly requires more computational effort to complete a run than both GREEDY and DSATUR. Indeed, as $n$ increases, this gap seems to widen quite significantly. In contrast, the GREEDY algorithm requires by far the fewest constraint checks, with its line being barely distinguishable with the horizontal axis in the figure.

The next graph, Figure 2.17, shows the computational requirements of the three algorithms for different values of $p$. It can be seen that the number of constraint checks required by GREEDY and DSATUR remains fairly stable over the range, suggesting that it is the number of vertices $n$, and not the edge connectivity $p$, that is the driving force in determining the two algorithms' computational requirements. In contrast, RLF's requirements once again increase quite rapidly over this range.

Finally, Figure 2.18 demonstrates the strong correlation that exists between the number of constraint checks the algorithms require and the subsequent CPU time that is used (coefficient of determination $R^2 = 0.939$).[2] In fact, the majority of this figure is once again dominated by data generated from runs of the RLF algorithm with GREEDY and DSATUR's results being tightly clustered in the bottom left corner (the GREEDY algorithm never required more than 16 ms on any of the graph colouring problem instances considered in this section; similarly the DSATUR algorithm never required more than 47 ms). This figure demonstrates that the use of constraint checks as a measure of computational effort is suitable for estimating CPU time, but also has the obvious advantage of being independent of any issues to do with computer hardware, programming languages and operating systems.

## 2.6 Chapter Summary and Further Reading

In this chapter we have reviewed a number of bounds for the graph colouring problem and have also compared and contrasted three constructive algorithms. For random graphs of different sizes and densities (including sets of graphs and their iso-

---

[2] The CPU times relate to a 3.0 GHz Windows 7 PC with 3.87 GB RAM.

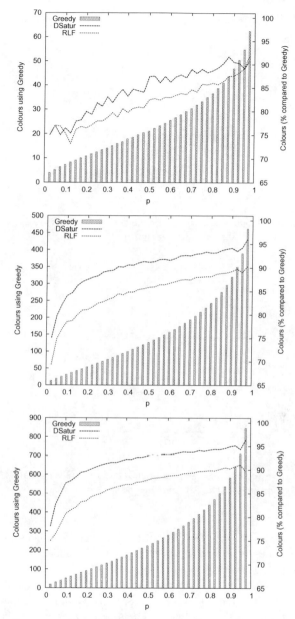

**Fig. 2.15** Mean quality of solutions achieved on random graphs $G_{n,p}$ by the RLF and DSATUR algorithms compared to GREEDY. All points are the mean across 50 graphs using $n = 100$, 1,000, and 2,000 respectively

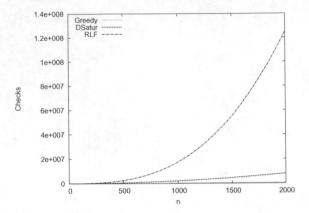

**Fig. 2.16** Number of constraint checks required by the GREEDY, DSATUR and RLF algorithms on random graphs $G_{n.p}$ with $p = 0.5$. All points are the mean of 50 trials

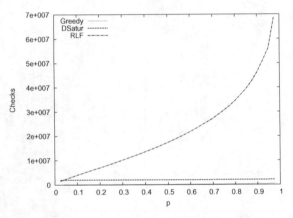

**Fig. 2.17** Number of constraint checks required by the GREEDY, DSATUR and RLF algorithms on random graphs $G_{n.p}$ with $n = 1,000$. All points are the mean of 50 trials

morphisms), we have seen that the RLF algorithm generally produces solutions with the fewest colours, though this also comes at the expense of added computation time, particularly for graphs with larger numbers of vertices.

In the next two chapters we will analyse a number of techniques that seek to improve upon the solutions produced by these algorithms. We now end this chapter by providing points of reference for further work on bounds for the chromatic number.

Reed (1999) has shown that Brooks' Theorem (2.7), can be improved by one colour when a graph $G$ has a sufficiently large value for $\Delta(G)$ and also has no cliques of size $\Delta(G)$. Specifically:

**Theorem 2.13 (Reed (1999))** *There exists some value $\delta$ such that if $\Delta(G) \geq \delta$ and $\omega(G) \leq \Delta(G) - 1$ then $\chi(G) \leq \Delta(G) - 1$.*

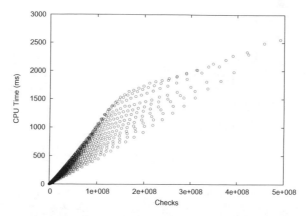

**Fig. 2.18** Scatter diagram showing the relationship between number of constraint checks and CPU time. Each point is the mean of the 50 runs performed by each algorithm on each $p$ and $n$ value

In the work a sufficient value for $\delta$ is shown to be $10^{14}$. Reed's Conjecture, also stated in this work, suggests that for any graph $G$,

$$\chi(G) \leq \left\lceil \frac{1 + \Delta(G) + \omega(G)}{2} \right\rceil. \tag{2.4}$$

A good survey on these issues can be found in the work of Cranston and Rabern (2014).

In Section 2.2.1.1 of this chapter we also saw that interval graphs (and more generally chordal graphs) feature chromatic numbers $\chi(G)$ equal to the their clique numbers $\omega(G)$. Chordal graphs form part of a larger family of graphs known as *perfect* graphs which, in addition to satisfying this criterion, are also known to maintain this property when any of its vertices are removed.

**Definition 2.16** A graph $G = (V, E)$ is perfect if, for every subgraph $G' \subseteq G$, $\chi(G') = \omega(G')$.

Defining the structures needed for a graph to be perfect has been the subject of much research in the field of graph theory and was eventually settled by Chudnovsky et al. (2006), who proved the earlier conjecture of Berge (1960), which states that a graph is perfect if and only if it contains no odd hole and no odd antihole. (A hole is an induced subgraph which is a cycle of length at least 4; an antihole is the complement). See MacKenzie (2002) for further details.

Looking at other topologies, bounds on the chromatic number of random graphs have also been determined by Bollobás (1988), who states that with very high probability, a random graph $G_{n,p}$ will have a chromatic number $\chi(G_{n,p})$:

$$\frac{n}{s} \leq \chi(G_{n,p}) \leq \frac{n}{s} + \left(1 + \frac{3 \log \log n}{\log n}\right) \tag{2.5}$$

where

$$s = \lceil 2\log_d n - \log_d \log_d n + 2\log_d(e/2) + 1 \rceil \qquad (2.6)$$

with $q = 1 - p$ and $d = 1/q$. The finding is based on calculating the expected number of disjoint cliques within a random graph, and provides much tighter bounds than Equation (2.3) and Brooks' Theorem (though recall that the latter applies to *all* graphs, not just random graphs). Further bounds on general graphs have also been given by Berge (1970), who finds

$$\frac{n^2}{n^2 - 2m} \leq \chi(G), \qquad (2.7)$$

and Hoffman (1970), who has shown

$$1 - \frac{\lambda_1(G)}{\lambda_n(G)} \leq \chi(G), \qquad (2.8)$$

where $\lambda_1(G)$ and $\lambda_n(G)$ are the biggest and smallest eigenvalues of the adjacency matrix of $G$. Both of these often give very loose lower bounds in practice, however.

Note that, strictly speaking, the three constructive algorithms reviewed in this section should be classed as heuristic algorithms as opposed to approximation algorithms. Unlike heuristics, approximation algorithms are usually associated with provable bounds on the quality of solutions they produce compared to the optimal. So for the graph colouring problem, using $A(G)$ to denote the number of colours used in a feasible solution produced by algorithm $A$ with graph $G$, a good approximation algorithm should feature an approximation ratio $A(G)/\chi(G)$ as close to 1 as possible. Those seeking an algorithm with a low approximation ratio for the graph colouring problem, however, should take note of the following theorem:

**Theorem 2.14 (Garey and Johnson (1976))** *If, for some constant $r < 2$ and constant $d$, there exists a polynomial-time graph colouring algorithm $A$ which guarantees to produce $A(G) \leq r \times \chi(G) + d$, then there also exists an algorithm $A'$ which guarantees $A'(G) = \chi(G)$.*

In other words, this states that we cannot hope to find an approximation algorithm $A$ for the graph colouring problem that, for all graphs, produces $A(G) < 2 \times \chi(G)$ unless P = NP.

# Chapter 3
# Advanced Techniques for Graph Colouring

In this chapter we now review many of the algorithmic techniques that can be (and have been) used for the graph colouring problem. The intention is to give the reader an overview of the different strategies available, including both exact and inexact methods. As we will see, a variety of different approaches are available, including backtracking algorithms, integer programming, evolutionary algorithms, neighbourhood search algorithms, and other metaheuristics. Full descriptions of these different techniques are provided as they arise in the text.

## 3.1 Exact Algorithms

Exact algorithms are those that will always determine the optimal solution to a computational problem, given excess time. One way of exactly solving NP-complete problems such as the graph colouring problem is to exhaustively search the solution space; however, as highlighted in Chapter 1, as problem sizes grow, the running time for exhaustive search soon becomes restrictively large. That said, it is still possible to design exact algorithms that are significantly faster than exhaustive search, though still not operating in polynomial time. Two such approaches are now considered.

### 3.1.1 Backtracking Approaches

We saw in Chapter 2 that algorithms such as GREEDY, DSATUR, and RLF can be used to quickly build feasible, though not necessarily optimal, solutions to instances of the graph colouring problem. It is also possible to extend these constructive algorithms to form what are known as *backtracking* algorithms. Backtracking is a general methodology that can be used for determining an optimal solution (or possibly all optimal solutions) to a computational problem such as graph colouring. In essence backtracking algorithms work by systematically building up partial solu-

© Springer International Publishing Switzerland 2016
R.M.R. Lewis, *A Guide to Graph Colouring*,
DOI 10.1007/978-3-319-25730-3_3

tions into complete solutions. However, during this construction process as soon as evidence is gained telling us that there is no way of completing the current partial solution to gain an optimal solution, the algorithm *backtracks* in order to try to find suitable ways of adjusting the current partial solution.

A simple backtracking algorithm for graph colouring might operate as follows. Given a graph $G = (V, E)$, the vertices are first ordered in some way so that vertex $v_i$ $(1 \le i \le n)$ corresponds to the $i$th vertex in this ordering. We also need to choose a value $k$ denoting the number of available colours. Initially this might be simply $k = \infty$. The backtracking algorithm then performs a series of *forward* and *backward* steps. Forward steps colour the vertices in the given order until a vertex is identified that cannot be feasibly coloured with one of the available $k$ colours. Backward steps on the other hand go back through the coloured vertices in reverse order and identify points where different colour assignments to vertices can be made. Forward steps are then resumed from these points. If a complete feasible colouring is found, then $k$ can be set to the number of colours used in this colouring minus 1, with the algorithm then continuing. Ultimately the algorithm terminates when a backward step reaches the root vertex $v_1$, or when some other stopping criterion such as a maximum time limit is met.

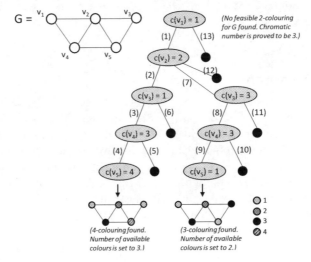

**Fig. 3.1** Example run of a backtracking algorithm

Figure 3.1 illustrates an example run of a simple backtracking algorithm on a small graph. Here the vertices are considered according to the sequence $v_1, v_2, \dots, v_5$. Vertices are also coloured with the lowest available feasible colour in the same fashion as in the GREEDY algorithm. As can be seen, the sequence of steps that the backtracking algorithm follows can be depicted as a tree. Each grey node in this tree represents a decision (an assignment of a vertex to a colour), and grey leaf nodes give a feasible solution. For clarity, the order in which the decision nodes are visited is shown by the numbers next to the corresponding edges in the tree. This corresponds to a depth first search of the tree.

In Steps (1) to (4) of Figure 3.1 we see that, following the behaviour of the GREEDY algorithm, all vertices are assigned to the feasible colour with the lowest index. The resultant solution is a 4-colouring, as shown in the figure. At this point we are now interested in finding a solution with fewer colours; hence the number of available colours $k$ is set to three. In Step (5) the algorithm has performed a backtrack and now tries to assign vertex $v_5$ to a new colour. However, colours 1, 2, and 3 are all infeasible for $v_5$; hence no further branches of the tree need to be explored from this node. This is signified by the black node in the search tree. Similarly, in Step (6) the algorithm has backtracked further in order to find a new colour for vertex $v_4$. However, colour 3 has already been tried, and colours 1 and 2 are not feasible so, again, no further branching from this node is required.

In the next step, Step (7), a new feasible colour is sought for vertex $v_3$. Colour 1 has already been tried, colour 2 is infeasible, but colour 3 is feasible. Hence $v_3$ is assigned to colour 3, and further forward steps are then carried out until a feasible 3-colouring of $G$ is achieved. At this point, the number of available colours is lowered to two, and the algorithm continues. According to the tree, when only two colours are available, there remain no other colour options for any of the previous vertices, so the algorithm backtracks all the way to the root of the tree, at which point it can terminate with the knowledge that no 2-colouring is available. This proves that the chromatic number of the graph is 3.

Of course, due to the NP-hardness of graph colouring, the time requirements of this sort of approach will often be excessively large, and executions will therefore need to be terminated prematurely, perhaps leaving the user with a suboptimal solution. On the other hand, the systematic construction of different solutions, together with the way in which the algorithm is able to ignore large swathes of the solution space (indicated by the black nodes in Figure 3.1), means that backtracking is usually far more efficient than approaches attempting a brute force enumeration of all candidate solutions.

Kubale and Jackowski (1985) have reviewed a number of ways in which backtracking algorithms for graph colouring might be enhanced through the addition of various heuristics, thereby encouraging good solutions to be encountered earlier in the search tree. These heuristics might include:

- Ordering the vertices in decreasing order of degree;
- Ordering vertices such that those with the fewest available colours are coloured first (in a similar fashion to that of the DSATUR algorithm);
- Assigning vertices to the largest colour classes first (with the rationale that forming large colour classes may lead to an overall reduction in the number of colour classes); and similarly
- Prioritising colour classes that contain large numbers of high-degree vertices.

A backtracking algorithm using the most effective heuristics seen in the survey of Kubale and Jackowski (1985) is considered further in Chapter 4.

## 3.1.2 Integer Programming

A second way of achieving an exact algorithm for graph colouring is to use integer programming (IP), which is a specialised type of linear programming model. Linear programming can be considered a general methodology for achieving optimal solutions to linear mathematical models. Such models can be said to consist of variables, linear constraints, and a linear objective function. The variables (or decisions) take on numerical values, and the constraints are used to define feasible ranges of values for these variables. The objective function is then used to measure the quality of a solution and to define which particular assignment of values to variables is considered optimal.

In general, the decision variables of linear programming models are continuous in the sense that they are permitted to be fractional. On the other hand, *integer* programming problems are a special type of linear programming model in which some, if not all, of the decision variables are restricted to integer values. Though this might seem like a subtle restriction, this insistence on integer-valued decision variables greatly increases the number of problems that can be modelled. Indeed, IP models can be used in a wide variety of applications, including supply chain design, resource management, timetabling, employee rostering and, as we shall see presently, the colouring of graphs.

Whereas algorithms such as the well-known simplex method are known to be effective for solving linear programs, there is no single preferred technique for solving integer programs. Instead a number of exact methods have been proposed, including branch-and-bound, cutting-plane, branch-and-price, as well as various hybrid techniques. Because of their wide applicability, a number of off-the-shelf software applications have also been developed in recent decades for solving linear and integer programming models, including commercial packages such as Xpress, CPLEX, and AIMMS, and free open source applications such as the Scip Optimisation suite (scip.zib.de) and Coin-OR (coin-or.org). Such packages allow users to input their particular model (in terms of variables, constraints, and objective function) and then simply click a button, at which point the software goes on to produce solutions using the exact methods just mentioned.

It is not our intention here to provide an in depth analysis of methods for solving IP models; instead we choose to focus on the ways in which the graph colouring problem can be formulated using IP principles. Readers interested in finding out more on the former are invited to consult the textbook of Wolsey (1998), which provides a thorough overview of the subject. Before considering graph colouring IP formulations however, the following points must be made:

- IP solution techniques such as branch-and-bound typically operate by subdividing the integer-valued region of the solution space in order to develop and refine bounds on the optimal cost. For problems where the aim is to minimise the objective function, an upper bound is defined as the cost of the best integer-valued solution found so far, while the lower bound is given by the optimal value of the associated linear programming model (that is, the IP formulation without the requirement of the decision variables being integer-valued). As a run of the algorithm progresses, these bounds will be improved until, eventually, an optimal solution is determined.

- There are often many different ways in which a particular problem can be formulated as an IP model. The task of finding a "good" formulation that can be solved effectively via techniques such as branch-and-bound is very important but might often be quite empirical in nature. However, avoiding features such as "symmetry" in the formulation will often help, as we shall see below.
- Many NP-hard problems, including graph colouring itself, can be formulated using integer programming; hence IP is an NP-hard problem in general. This implies that exact techniques such as branch-and-bound and cutting-planes will often be unable to determine optimal solutions in reasonable time. Indeed, in some cases the algorithms might even fail to determine any integer valued solution in acceptable run time, meaning the user will be left with no valid solution with which to work.

Let us now consider some IP formulations of the graph colouring problem. As usual, let $G = (V, E)$ be a graph with $n$ vertices and $m$ edges. Perhaps the most simple formulation involves using two binary matrices $\mathbf{X}_{n \times n}$ and $\mathbf{Y}_n$ for holding the variables of the problem. These are interpreted as follows:

$$X_{ij} = \begin{cases} 1 \text{ if vertex } v_i \text{ is assigned to colour } j, \\ 0 \text{ otherwise.} \end{cases} \tag{3.1}$$

$$Y_j = \begin{cases} 1 \text{ if at least one vertex is assigned to colour } j, \\ 0 \text{ otherwise.} \end{cases} \tag{3.2}$$

Note that the elements of $\mathbf{X}$ and $\mathbf{Y}$ are required not only to be integers here, but also binary. This is a common restriction in integer programming models and is necessary here because only two options are available: vertex $v_i$ is either assigned to colour $j$, or it is not. The objective in this model is to now minimise the number of colours being used according to the objective function

$$\min \sum_{j=1}^{n} Y_j \tag{3.3}$$

subject to the following constraints being satisfied:

$$X_{ij} + X_{lj} \leq Y_j \quad \forall \{v_i, v_l\} \in E, \ \forall j \in \{1, \ldots, n\} \tag{3.4}$$

$$\sum_{j=1}^{n} X_{ij} = 1 \quad \forall v_i \in V. \tag{3.5}$$

Here, Constraint (3.4) ensures that no pair of adjacent vertices are assigned to the same colour and that $Y_j = 1$ if and only if some vertex is assigned to colour $j$. Constraint (3.5) then specifies that each vertex should be assigned to exactly one colour.

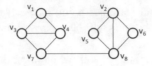

**Fig. 3.2** Example graph with $n = 8$ vertices and $m = 12$ edges

An implementation of this model in the Xpress-Mosel language together with ex-ample input and output is given in Appendix A.3. To illustrate this model, consider the small graph given in Figure 3.2, which comprises $n = 8$ vertices and $m = 12$ edges. On running the Xpress optimisation algorithm (which utilises a standard branch-and-bound framework), the following optimal solution is returned:

$$\mathbf{X} = \begin{pmatrix} 1\,0\,0\,0\,0\,0\,0\,0 \\ 0\,1\,0\,0\,0\,0\,0\,0 \\ 0\,0\,0\,0\,0\,0\,1\,0 \\ 0\,1\,0\,0\,0\,0\,0\,0 \\ 1\,0\,0\,0\,0\,0\,0\,0 \\ 1\,0\,0\,0\,0\,0\,0\,0 \\ 1\,0\,0\,0\,0\,0\,0\,0 \\ 0\,0\,0\,0\,0\,0\,1\,0 \end{pmatrix}$$

$$\mathbf{Y} = \begin{pmatrix} 1\,1\,0\,0\,0\,0\,1\,0 \end{pmatrix}$$

As can be seen, the cost of this optimal solution is $\sum_{j=1}^{n} Y_j = 3$, which is the chro-matic number for this graph. Specifically, the colours labelled 1, 2 and 7 are assigned to the vertices as follows: $c(v_1) = 1$, $c(v_2) = 2$, $c(v_3) = 7$, $c(v_4) = 2$, $c(v_5) = 1$, $c(v_6) = 1$, $c(v_7) = 1$, and $c(v_8) = 7$.

If we now remove the requirement that the elements in $\mathbf{X}$ need to be binary, we get the following optimal solution to the associated linear programming problem:

$$\mathbf{X} = \begin{pmatrix} 0\,0.5\,0\,0\,0\,0\,0.5\,0 \\ 0\,0.5\,0\,0\,0\,0\,0.5\,0 \\ 0\,0.5\,0\,0\,0\,0\,0.5\,0 \\ 0\,0.5\,0\,0\,0\,0\,0.5\,0 \\ 0\,0.5\,0\,0\,0\,0\,0.5\,0 \\ 0\,0.5\,0\,0\,0\,0\,0.5\,0 \\ 0\,0.5\,0\,0\,0\,0\,0.5\,0 \\ 0\,0.5\,0\,0\,0\,0\,0.5\,0 \end{pmatrix}$$

$$\mathbf{Y} = \begin{pmatrix} 0\,1\,0\,0\,0\,1\,0 \end{pmatrix}$$

The cost of this solution is 2, but it is obviously not valid from the point of view of graph colouring because each vertex is specified as being (nonsensically) "half assigned" to colour 2 and "half assigned" to colour 7. However, this relaxation does provide us with a lower bound to the problem, telling us that at least two colours are needed in any feasible solution.

One of the major drawbacks with the above IP model is its underlying symmetry. Indeed, it can be observed that any feasible $k$-colouring of a graph can be expressed in $^nP_k = \frac{n!}{(n-k)!}$ ways by simply permuting the columns of the $\mathbf{X}$ matrix. This has the potential to make any IP optimisation algorithm very inefficient as it can drastically increase the number of solutions that it needs to consider in order to find and prove optimality. One way to alleviate this problem might be to introduce a further constraint,

$$Y_j \geq Y_{j+1} \quad \forall j \in \{1,\ldots,n-1\}, \tag{3.6}$$

which ensures that any $k$-coloured solution only uses the colours labelled 1 to $k$. Using this constraint, an optimal solution to our current problem is now

$$\mathbf{X} = \begin{pmatrix} 1 & 0 & 0 & 0 & 0 & 0 & 0 & 0 \\ 0 & 0 & 1 & 0 & 0 & 0 & 0 & 0 \\ 0 & 0 & 1 & 0 & 0 & 0 & 0 & 0 \\ 0 & 1 & 0 & 0 & 0 & 0 & 0 & 0 \\ 1 & 0 & 0 & 0 & 0 & 0 & 0 & 0 \\ 1 & 0 & 0 & 0 & 0 & 0 & 0 & 0 \\ 1 & 0 & 0 & 0 & 0 & 0 & 0 & 0 \\ 0 & 1 & 0 & 0 & 0 & 0 & 0 & 0 \end{pmatrix}$$

$$\mathbf{Y} = \begin{pmatrix} 1 & 1 & 1 & 0 & 0 & 0 & 0 & 0 \end{pmatrix}$$

which, as shown, only uses colours 1, 2, and 3. Observe that this partition of the vertices is different from that of the previous case, though it is still optimal. In this case the algorithm has simply returned the first optimal solution it has encountered, which just happens to be different under Constraint (3.6). Note, however, that although this extra constraint provides a significant improvement on the first model, there are still equivalent solutions that arise due to the $k!$ ways in which the first $k$ colours can be permuted. To eliminate some of these equivalent solutions we might instead choose to replace Constraint (3.6) with the following:

$$\sum_{i=1}^{n} X_{ij} \geq \sum_{i=1}^{n} X_{ij+1} \quad \forall j \in \{1,\ldots,n-1\}. \tag{3.7}$$

This ensures that the number of vertices assigned to colour $j$ is always greater than or equal to the number of vertices assigned to colour $j+1$. This again presents improvements over the preceding model, but there still exists some symmetry due to colour classes of the same size being interchangeable. One final improvement suggested by Méndez-Díaz and Zabala (2008) might therefore be obtained by replacing Constraint (3.7) with the following two constraints:

$$X_{ij} = 0 \quad \forall v_i \in V, \; j \in \{i+1,\ldots,n\} \tag{3.8}$$

$$X_{ij} \leq \sum_{l=j-1}^{i-1} X_{lj-1} \quad \forall v_i \in V - \{v_1\}, \; \forall j \in \{2,\ldots,i-1\}. \tag{3.9}$$

Together, these constraints specify a unique permutation of the first $k$ columns for each possible $k$-colouring. Specifically, vertex $v_1$ must be assigned to colour 1, $v_2$ must be assigned to either colour 1 or colour 2, and so on. (Or in other words, the columns are sorted according to the minimally labelled vertex in each colour class.) Under these constraints the optimal solution to our example problem is now:

$$\mathbf{X} = \begin{pmatrix} 1\,0\,0\,0\,0\,0\,0\,0 \\ 0\,1\,0\,0\,0\,0\,0\,0 \\ 0\,1\,0\,0\,0\,0\,0\,0 \\ 0\,0\,1\,0\,0\,0\,0\,0 \\ 1\,0\,0\,0\,0\,0\,0\,0 \\ 1\,0\,0\,0\,0\,0\,0\,0 \\ 1\,0\,0\,0\,0\,0\,0\,0 \\ 0\,0\,1\,0\,0\,0\,0\,0 \end{pmatrix}$$

$$\mathbf{Y} = \begin{pmatrix} 1\,1\,1\,0\,0\,0\,0\,0 \end{pmatrix}$$

as required.

Obviously, although a model with reduced symmetry might lead to improved solution times, we must also pay heed to other features of the model such as the number of variables and constraints, and the performance of our chosen solver on the associated linear programming relaxation. As mentioned earlier, the effectiveness of an IP formulation for a specific problem will often only be confirmed after empirical observation.

### 3.1.2.1  Independent Set Selection

An alternative IP formulation for the graph colouring problem involves considering the independent sets of a graph. Suppose, given a graph $G$, we were able to form a complete list of all of its independent sets $\mathcal{I} = \{I_1, I_2 \ldots, I_l\}$. Given this set, we could then define a binary vector $\mathbf{Y}_l$ such that

$$Y_j = \begin{cases} 1 & \text{if independent set } I_j \text{ is included in the solution,} \\ 0 & \text{otherwise,} \end{cases} \tag{3.10}$$

with the objective function

$$\min \sum_{j=1}^{l} Y_j \tag{3.11}$$

and the single constraint

$$\sum_{j:v_i \in I_j} Y_j = 1 \ \forall v_i \in V. \tag{3.12}$$

In this formulation we are therefore concerned with selecting a subset of independent sets from $\mathcal{I}$ so that each vertex of the graph is included in exactly one of these independent sets (Constraint (3.12)), with the number of selected independents sets being minimised. The selected independent sets then correspond to colour classes, and a feasible solution minimising the objective function corresponds to an optimal colouring.

Such a formulation is both elegant and avoids any issues of symmetry. Unfortunately, however, it is impractical for most graphs because the number of different independent sets grows exponentially in relation to the number of vertices, leading to a formulation that is far too large to compute. One way of dealing with this feature is to make use of variable generation, whereby decision variables (in this case the independent sets) are only generated as and when they are needed. A typical approach might operate by first generating a relatively small number of independent sets, perhaps via heuristics, and then solving the linear relaxation of the formulation. Information might then be gleaned from this solution as to whether there exist independent sets that are not included in the model, but whose addition has the potential to improve the solution. If such sets are seen to exist, these can be generated (again perhaps via heuristics) and then added to the formulation. This process might then be used in conjunction with branch-and-bound (or something similar) until an acceptable integer solution is obtained.

IP formulations and associated algorithms for the graph colouring problem have been presented by, amongst others, Mehrotra and Trick (1996), Hansen et al. (2009), and Malaguti et al. (2011). As with backtracking algorithms, a distinct advantage of using IP is that the associated algorithms are complete and can therefore prove the optimality of their returned solutions given excess time. In addition IP can also often return good lower bounds on the chromatic number for many graphs. In practice however, these advantages are often only apparent with smaller graphs or those with certain types of topologies.

## 3.2 Inexact Heuristics and Metaheuristics

Perhaps the most active and fruitful avenue in the design of algorithms for graph colouring in recent decades has been in the application of heuristic- and metaheuristic-based approaches. In contrast to backtracking and IP methods, these approaches are typically *inexact* and as such are not guaranteed to return an optimal solution, even if granted excess time. Worse, even if they do happen to achieve an optimal solution, these methods will often fail to recognise this and will continue to execute until some user-defined stopping criterion is met.

On the other hand, heuristics and metaheuristics have the advantage of being highly adaptable to different problems. As a result, various algorithms of this type have been shown to produce excellent results on a large range of different graph colouring problems. Often heuristics and metaheuristics are also considered to produce better solutions than exact algorithms for larger instances, though this depends on the makeup of the graph, together with the amount of time the user is willing to wait. An oft-quoted rule of thumb in the literature seems to be that heuristic algorithms are considered to be more useful than exact graphs whenever $n$ is roughly larger than 100 (see for example Galinier and Hertz (2006)); however, there are many exceptions to this rule, as we will see in Chapter 4.

In this section we survey a number of these algorithms. In our descriptions we will often make reference to different types of metaheuristics and will describe these in more detail as they arise in the text. For now, it suffices to say that metaheuristics can be considered types of *higher level* algorithmic frameworks that are intended to be applicable to a variety of different optimisation problems with relatively few modifications needing to be made in each case. Examples of metaheuristics include evolutionary algorithms, simulated annealing, tabu search, and ant colony optimisation. Typically they operate by navigating their way through a large set of candidate solutions (the *solution space*), attempting to optimise an objective function that reflects each member's quality. In some cases metaheuristics will also attempt to make educated decisions on which parts of the solution space to explore next based on the characteristics of candidate solutions already observed. This allows them to "learn" about the solution space during a run and therefore hopefully make good decisions about where the best solutions can be found.

In terms of the graph colouring problem, it is helpful to classify heuristic and metaheuristic methods according to how their solution spaces are defined. Specifically, we can consider feasible-only solution spaces, spaces of complete improper $k$-colourings, and spaces of incomplete proper $k$-colourings. Each of these categories is now considered in turn.

## 3.2.1 Feasible-Only Solution Spaces

The first set of algorithms we consider are those that operate within spaces of feasible colourings. Approaches of this type seek to identify solutions within this space that feature small numbers of colours. Often these methods make use of the GREEDY algorithm to construct solutions; hence, they are concerned with identifying good permutations of the vertices. (Recall from Theorem 2.2 that, for any graph, a permutation of the vertices always exists that decodes into an optimal solution via the application of GREEDY.)

One early example of this type of approach was the *iterated greedy* algorithm of Culberson and Luo (1996). This rather elegant algorithm exploits the findings of Theorem 2.1: namely that, given a feasible colouring $S$, a permutation of the vertices can be formed that, when fed back into the GREEDY algorithm, results in a new solution $S'$ that uses equal or fewer colours than $S$. To start, DSATUR is used to produce an initial feasible solution. Then, at each iteration, the current solution $S = \{S_1, \ldots, S_{|S|}\}$ is taken and its colour classes are reordered to form a new permutation of the vertices. This permutation is then used with GREEDY to produce a new feasible solution before the process is repeated indefinitely.

Culberson and Luo (1996) suggest a number of ways in which reorderings of the colour classes can be achieved at each iteration. These include:

- Largest First: Arranging the colour classes in order of decreasing size;
- Reverse: Reversing the order of the colour classes in the current solution; and
- Random: Rearranging the colour classes randomly.

The Largest First heuristic is used in an attempt to construct large independent sets in the graph, while the Reverse heuristic encourages vertices to be mixed among different colour classes. The Random heuristic is then used to prevent the algorithm from cycling, allowing new regions of the solution space to be explored. Culberson and Luo (1996) ultimately recommend selecting these heuristics randomly at each iteration according to the ratio 5:5:3 respectively.

Two other algorithms operating within this solution space are the evolutionary algorithms (EAs) of Mumford (2006) and Erben (2001). EAs are a type of metaheuristic inspired by biological evolution and they operate by maintaining a *population* of candidate solutions representing a small sample of the solution space. During a run, EAs attempt to improve the quality of members of the population using the following operators:

Recombination. This aims to create new solutions by combining different parts of existing members of the population. Often these "parts" are referred to as "building blocks", and the existing and new solutions are known as "parent" and "offspring" solutions respectively;

Mutation. This makes changes to a candidate solution in order to allow new regions of the solution space to be explored. These changes might be made at random, or perhaps via some sort of local search or improvement operator.

Evolutionary Pressure. As with biological evolution, EAs usually also exhibit some bias towards keeping good candidate solutions in the population, and rejecting bad ones. Hence, high-quality solutions might be more likely to be used for recombination, and weaker solutions might be more susceptible to being replaced in the population by newly created offspring solutions.

The evolution, and hopefully improvement, of an EA's population takes place with the repeated application of the above operators; however, it is often necessary to design specialised recombination and mutation operators that can suitably exploit the underlying structures of the problem at hand. The recombination operator of Erben (2001) seeks to do this by considering colour classes as the underlying building blocks of the graph colouring problem. Specifically, their recombination operates by first taking two feasible parent solutions $S_1$ and $S_2$. A subset of colour classes are then selected from $S_2$ and copied into a copy of $S_1$ to form a new offspring solution, say $S'$. At this point, $S'$ will contain multiple occurrences of some vertices, and so the algorithm goes through the colour classes of $S'$ and deletes all colour classes containing a duplicate that came from parent $S_1$. This operation results in an offspring solution that is proper, but most likely partial; thus any missing vertices are randomly permuted and reinserted back into $S'$ using the GREEDY algorithm to form a feasible solution. An example of this process is shown in Figure 3.3.

One notable feature of this recombination operator is that, before the GREEDY algorithm is used to reinsert missing vertices, each colour class in the offspring will be a copy of a colour class existing in at least one of the parents. That is:

| | Parent $S_1$ | Parent $S_2$ | Offspring $S'$ | Comments |
|---|---|---|---|---|
| 1) | $\{\{v_1,v_2,v_3\},$ $\{v_4,v_5,v_6,v_7\},$ $\{v_8,v_9,v_{10}\}\}$ | $\{\{v_3,v_4,v_5,v_7\},$ $\{v_1,v_6,v_9\},$ $\{v_2,v_8\},$ $\{v_{10}\}\}$ | $\{\{v_1,v_2,v_3\},$ $\{v_4,v_5,v_6,v_7\},$ $\{v_8,v_9,v_{10}\}\}$ | $S'$ is initially a copy of Parent $S_1$ |
| 2) | $\{\{v_1,v_2,v_3\},$ $\{v_4,v_5,v_6,v_7\},$ $\{v_8,v_9,v_{10}\}\}$ | $\{\{v_3,v_4,v_5,v_7\},$ $\{v_1,v_6,v_9\},$ $\{v_2,v_8\},$ $\{v_{10}\}\}$ | $\{\{v_1,v_2,v_3\},$ $\{v_4,v_5,v_6,v_7\},$ $\{v_8,v_9,v_{10}\},$ $\{v_2,v_8\},$ $\{v_{10}\}\}$ | Randomly select colour classes from $S_2$ and copy them to $S'$ |
| 3) | $\{\{v_1,v_2,v_3\},$ $\{v_4,v_5,v_6,v_7\},$ $\{v_8,v_9,v_{10}\}\}$ | $\{\{v_3,v_4,v_5,v_7\},$ $\{v_1,v_6,v_9\},$ $\{v_2,v_8\},$ $\{v_{10}\}\}$ | $\{\{v_4,v_5,v_6,v_7\},$ $\{v_2,v_8\},$ $\{v_{10}\}\}$ | Remove classes from $S'$ that came from $S_1$ and that contain duplicates. In this case vertices $v_1$, $v_3$ and $v_9$ and are now missing from $S'$ |
| 4) | $\{\{v_1,v_2,v_3\},$ $\{v_4,v_5,v_6,v_7\},$ $\{v_8,v_9,v_{10}\}\}$ | $\{\{v_3,v_4,v_5,v_7\},$ $\{v_1,v_6,v_9\},$ $\{v_2,v_8\},$ $\{v_{10}\}\}$ | $\{\{v_4,v_5,v_6,v_7\},$ $\{v_2,v_8,v_9,v_1\},$ $\{v_{10},v_3\}\}$ | Reinsert missing vertices into $S'$ using GREEDY to form a feasible colouring. |

**Fig. 3.3** Example application of the recombination operator of Erben (2001)

$$S_i \in S' \implies S_i \in (S_1 \cup S_2). \tag{3.13}$$

In particular, if a colour class is seen to exist in both parents then this colour class will be guaranteed to be present in the offspring solution. Features such as these are generally considered desirable in a recombination operator in that they provide a mechanism by which building blocks (colour classes in this case) can be passed from parents to offspring. For this particular operator however, there is also the possibility that an offspring might inherit all of its colour classes from the second parent due to the policy of deleting colour classes originating from the first parent.

The mutation operator of this algorithm works in a similar fashion to recombination by deleting some randomly selected colour classes from a solution, randomly permuting these vertices, and then reinserting them into the solution via GREEDY.

In his EA for graph colouring, Erben also proposes the following heuristic-based objective function:

$$f_1(S) = \frac{\sum_{S_i \in S} \left( \sum_{v \in S_i} \deg(v) \right)^2}{|S|} \tag{3.14}$$

where $\sum_{v \in S_i} \deg(v)$ gives the total degree of all vertices assigned to the colour class $S_i$. Here, the aim is to maximise $f_1$ by making increases to the numerator (by forming large colour classes in which high-degree vertices are grouped together) and decreases to the denominator (by reducing the number of colour classes). It is also suggested that this objective function allows evolutionary pressure to be sustained in a population for longer during a run compared to the more obvious choice of using the number of colours $|S|$, because it allows greater distinction between individuals.

The EA of Mumford (2006) also seeks to construct offspring solutions by combining the colour classes of parent solutions. In her research, two recombination operators are suggested, the Merge Independent Sets (MIS) operator and the Permutation One Point (POP) operator. The MIS operator starts by taking two feasible parent solutions, $S_1$ and $S_2$, and constructing two permutations. As with the Iterated Greedy algorithm, vertices within the same colour classes are put into adjacent positions in these permutations. For example, the two solutions

$$S_1 = \{\{v_1, v_2, v_3\}, \{v_4, v_5, v_6, v_7\}, \{v_8, v_9, v_{10}\}\}$$
$$S_2 = \{\{v_1, v_6, v_9\}, \{v_2, v_8\}, \{v_3, v_4, v_5, v_7\}, \{v_{10}\}\}$$

might result in the following two vertex permutations:

$$\pi^{(1)} = (v_1, v_2, v_3 : v_4, v_5, v_6, v_7 : v_8, v_9, v_{10})$$
$$\pi^{(2)} = (\mathbf{v_1, v_6, v_9} : \mathbf{v_2, v_8} : \mathbf{v_3, v_4, v_5, v_7} : \mathbf{v_{10}}).$$

(For convenience, colons are used in these permutations to mark boundaries between different colour classes). In the next step of the operator, the two permutations are merged randomly such that the boundaries between the colour classes are maintained. For example, we might merge the above examples to get:

$$(\mathbf{v_1, v_6, v_9} : v_1, v_2, v_3 : v_4, v_5, v_6, v_7 : \mathbf{v_2, v_8} : \mathbf{v_3, v_4, v_5, v_7} : v_8, v_9, v_{10} : \mathbf{v_{10}}).$$

Finally, two offspring permutations are then formed by using the first occurrence of each vertex for the first offspring, and the second occurrence for the second offspring:

$$\pi'^{(1)} = (\mathbf{v_1, v_6, v_9}, v_2, v_3, v_4, v_5, v_7, \mathbf{v_8}, v_{10})$$
$$\pi'^{(2)} = (v_1, v_6, \mathbf{v_2, v_3, v_4, v_5, v_7}, v_8, v_9, \mathbf{v_{10}}).$$

These two permutations are then converted into two feasible offspring solutions by feeding them into the GREEDY algorithm.

The POP operator of Mumford (2006) follows a similar scheme by first forming two permutations, $\pi^{(1)}$ and $\pi^{(2)}$, as above. A random cut point is then chosen, and the first portion of $\pi^{(1)}$ up to the cut point becomes the first portion of the second offspring. The remainder of the second offspring is then obtained by copying the vertices absent from the first portion of the offspring in the same sequence as they occur in the second parent $\pi^{(2)}$. The first offspring is found in the same way, but with the roles of the parents reversed. For example, using "|" to signify the cut point, the permutations

$$\pi^{(1)} = (v_1, v_2, v_3, v_4 \mid v_5, v_6, v_7, v_8, v_9, v_{10})$$
$$\pi^{(2)} = (\mathbf{v_1, v_6, v_9, v_2} \mid \mathbf{v_8, v_3, v_4, v_5, v_7, v_{10}})$$

result in the following new permutations:

$$\pi'^{(1)} = (v_1, v_2, v_3, v_4, \mathbf{v_6, v_9, v_8, v_5, v_7, v_{10}}) \text{ and}$$
$$\pi'^{(2)} = (\mathbf{v_1, v_6, v_9, v_2}, v_3, v_4, v_5, v_7, v_8, v_{10}).$$

As before, two offspring solutions are then formed by feeding these new permutations into the GREEDY algorithm.

As we have seen, the recombination operators used in the EAs of both Erben (2001) and Mumford (2006) attempt to provide mechanisms by which good colour classes within a population can be propagated, thereby hopefully allowing good offspring solutions to be formed. However, it is evident that the overall performance of their algorithms as presented in their papers does not seem as strong as that of other algorithms reported in the literature. That said, as we shall see later some of the best results in the field have been actually been produced when evolutionary-based algorithms have been hybridised with more aggressive local search-based procedures. Future research might also determine whether this is also the case with these operators.

Moving away from EAs, the technique of Lewis (2009) also operates in the space of feasible solutions. In this case their algorithm makes use of operators based on the Iterated Greedy algorithm for making large changes to a solution, and combines these with specialised local search operators for making smaller, refining changes. These are the so-called *Kempe chain interchange* and *pair swap* operators, which are defined as follows:

**Definition 3.1** *Let* $S = \{S_1, \ldots, S_k\}$ *be a feasible solution. Given an arbitrary vertex* $v \in S_i$ *and a second colour class* $S_j$ *(*$1 \leq i \neq j \leq k$*), a* Kempe chain *is defined as a connected subgraph that contains* $v$*, and that only comprises vertices coloured with colours* $i$ *and* $j$*. The set of vertices involved in such a chain is denoted by* KEMPE$(v, i, j)$*. A Kempe chain interchange* involves taking a particular Kempe chain and swapping the colours of all vertices contained within it.

**Definition 3.2** *Let the Kempe chains* KEMPE$(u, i, j)$ *and* KEMPE$(v, j, i)$ *both contain just one vertex each (therefore implying that* $u$ *and* $v$ *are nonadjacent.) A* pair swap *involves swapping the colours of* $u$ *and* $v$*.*

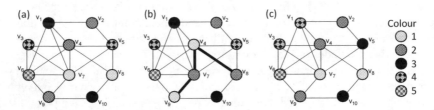

**Fig. 3.4** An example 5-colouring (a); the result of a Kempe chain interchange using KEMPE$(v_7, 1, 2)$ (b); and the result of a pair swap using $v_1$ and $v_5$ (c)

Figure 3.4 shows examples of these operators. In Figure 3.4(a) we see that KEMPE$(v_7, 1, 2) = \{v_4, v_7, v_8, v_9\}$. Interchanging the colours of these vertices gives the colouring shown in Figure 3.4(b). For an example pair swap, observe that in Figure 3.4(a) the Kempe chains identified by both KEMPE$(v_1, 3, 4)$ and KEMPE$(v_5, 4, 3)$ contain just one vertex each. Hence a pair swap can be performed, as is the case in Figure 3.4(c).

The fact that applications of these operators will always preserve the feasibility of a solution is due to the following theorem:

**Theorem 3.1** *Given a proper solution* $S = \{S_1, \ldots, S_k\}$*, the application of a Kempe chain interchange or a pair swap will result in a new solution* $S'$ *that is also proper.*

*Proof.* For the Kempe chain interchange operator, consider the situation where $S$ is proper but $S'$ is not. Because a Kempe chain interchange involves two colours $i$ and $j$, $S'$ must feature a pair of adjacent vertices $u$ and $v$ that are assigned the same colour. Without loss of generality, assume this to be colour $i$. Moreover, $u$ and $v$ must have both been in the Kempe chain used for the interchange since they are adjacent, implying $u$ and $v$ are both assigned to colour $j$ in $S$. However, this is impossible since $S$ is known to be proper.

According to the conditions given in Definition 3.2, $v$ cannot be adjacent to any vertex coloured with $j$, and $u$ cannot be adjacent to any vertex coloured with $i$. Hence swapping the colours of $u$ and $v$ will also ensure that $S'$ is proper. □

The method of Lewis (2009) has been shown to outperform those of both Erben (2001) and Culberson and Luo (1996) on a variety of different graphs. Consequently, this forms one of our case study algorithms discussed in Chapter 4.

## 3.2.2 Spaces of Complete, Improper k-Colourings

Perhaps the majority of metaheuristic algorithms proposed for graph colouring have been designed for exploring the space of complete improper colourings. Such methods typically start by proposing a fixed number of colours $k$, with each vertex then being assigned to one of these colours using heuristics, or possibly at random. During this assignment there may exist vertices that cannot be assigned to any colour without inducing a clash, but these will be assigned to one of the colours anyway. (Recall that a clash occurs when a pair of adjacent vertices are assigned to the same colour—see Definition 1.2.)

The above assignment process leaves us with a $k$-partition of the vertices that represents a complete, but most likely improper $k$-colouring. A natural way to measure the quality of this solution is to then simply count the number of clashes. This can be achieved via the following objective function:

$$f_2(S) = \sum_{\forall\{u,v\}\in E} g(u,v) \qquad (3.15)$$

where

$$g(u,v) = \begin{cases} 1 & \text{if } c(u) = c(v) \\ 0 & \text{otherwise.} \end{cases}$$

The aim of algorithms using this solution space is to make alterations to the $k$-partition so that the number of clashes is reduced to zero. If this is achieved, $k$ might then be reduced and the process restarted. Alternatively if all clashes cannot be eliminated, $k$ can be increased. Note that at each setting for $k$, we are thus attempting to solve the NP-complete decision variant of the graph colouring problem: "can the graph $G$ be feasibly coloured using $k$ colours?"

Perhaps the first algorithm to make use of the above strategy was due to Chams et al. (1987), who made use of the simulated annealing metaheuristic. Soon after this, Hertz and de Werra (1987) proposed a similar algorithm called TABUCOL based on the tabu search metaheuristic of Glover (1986). Simulated annealing and tabu search are types of metaheuristics based on the concept of local search (sometimes known as neighbourhood search). In essence local search algorithms make use of *neighbourhood operators* which are simple schemes for changing (or disrupting) a particular candidate solution. In the examples just cited, this operator is simply:

- Take a vertex $v$ currently assigned to colour $i$, and move it to a new colour $j$ (where $1 \leq i \neq j \leq k$).

Given a particular candidate solution $\mathcal{S}$, the *neighbourhood* of $\mathcal{S}$, denoted by $N(\mathcal{S})$, is then defined as the set of all candidate solutions that can be found from $\mathcal{S}$ via some application of this operator.

A very simple way of performing local search with such a neighbourhood operator is to make use of the elementary *random descent* method. This starts by generating an initial solution $S$ by some means, and then evaluating it according to the chosen objective function $f(\mathcal{S})$ which, for now, we assume we are seeking to minimise. At each iteration of the algorithm a *move* in the solution space is attempted by randomly applying the neighbourhood operator to the incumbent solution $\mathcal{S}$ to form a new solution $\mathcal{S}'$ (that is, the new solution $\mathcal{S}'$ is chosen randomly from the set $N(\mathcal{S})$). If this new solution is seen to be better than the incumbent (i.e., $f(\mathcal{S}') < f(\mathcal{S})$) then it set as the incumbent for the next iteration (i.e., $\mathcal{S} \leftarrow \mathcal{S}'$); otherwise no changes occur. The algorithm can then be left to run indefinitely or until some stopping criterion is met.

Though the random descent method is very intuitive, it is highly susceptible to getting caught at local optima within the solution space. This occurs when all neighbours of the incumbent solution feature an equal or inferior cost—that is, $\forall \mathcal{S}' \in N(\mathcal{S}), f(\mathcal{S}') \geq f(\mathcal{S})$. It is obvious that if a random descent algorithm reaches such a point in the solution space, then no further improvements (or changes to the solution) will be possible. The *simulated annealing* algorithm is a generalisation of random descent which offers a mechanism by which this issue might be circumnavigated. In essence, the main difference between the two methodologies lies in the criterion used for deciding whether to perform a move or not. As noted, for random descent this criterion is simply $f(\mathcal{S}') < f(\mathcal{S})$. Simulated annealing uses this criterion, but also accepts a move to a worse solution with probability $\exp(-\delta/t)$, where $\delta = |f(\mathcal{S}) - f(\mathcal{S}')|$ gives the proposed change in cost and $t$ is a parameter known as the *temperature*. Typically, in simulated annealing $t$ is set to a relatively high value at the beginning of execution. This results in nearly all moves in the solution space being accepted, meaning that the exploration method closely resembles a random walk. During a run $t$ is then slowly reduced, meaning that the chances of accepting a worsening move become increasingly less likely, causing the algorithm's behaviour to approach that of the random descent method. This additional acceptance criterion often allows the algorithm to escape from local optima, allowing SA to explore a greater span of the solution space compared to random descent.

A pseudocode description of an example simulated annealing algorithm is given in Figure 3.5. In this case the temperature $t$ is reduced every $z$ iterations by mul-

| SIMULATED-ANNEALING $(i \leftarrow 0)$ |
|---|
| (1) Produce an initial solution $S$ |
| (2) Choose an initial value for $t$ |
| (3) **while (not** stopping condition) **do** |
| (4)       Randomly choose $S' \in N(S)$ |
| (5)       **if** $f(S') \leq f(S)$ **then** |
| (6)          $S \leftarrow S'$ |
| (7)       **else if** $r \leq \exp(-\delta/t)$ **then** |
| (8)          $S \leftarrow S'$ |
| (9)       $i \leftarrow i+1$ |
| (10)      **if** $i \pmod z) = 0$ **then** |
| (11)         $t \leftarrow \alpha t$ |

**Fig. 3.5** The Simulated Annealing algorithm. Here, a random value for $r \in [0,1]$ is generated in each iteration. All other notation used here is described in the accompanying text

tiplying it by a *cooling rate* $\alpha \in (0,1)$. Many other cooling schemes are possible, however. Since its introduction by Kirkpatrick et al. (1983), simulated annealing has become a well-known and often very successful method for combinatorial optimisation problems, including applications in areas such as scheduling (Sekiner and Kurt, 2007), university timetabling (Lewis and Thompson, 2015), packing problems (Egeblad and Pisinger, 2009), and bridge construction (Perea et al., 2008). Methods based on simulated annealing were also winners in first two International Timetabling Competitions held in 2003 and 2007, where competitors were asked to design algorithms for producing timetables for a set of benchmark problem instances (see cs.qub.ac.uk/itc2007/ and Chapter 8).

One potentially problematic feature of the simulated annealing metaheuristic is that it does not maintain any memory of the solutions previously observed within the solution space. Indeed, it may often visit the same solution multiple times, or could even spend significant amounts of time cycling within the same subset of solutions. In contrast to this, the *tabu search* metaheuristic contains mechanisms that are intended to help avoid cycling, therefore encouraging the algorithm to enter new regions of the solution space.

In the same way that simulated annealing can be considered a generalisation of random descent, the tabu search algorithm can be seen as an extension of *steepest descent* methodology. Steepest descent acts in a similar fashion to random descent in that it starts with an initial solution $S$ and then repeatedly applies a neighbourhood operator to try to make improvements. In contrast however, at each iteration of the steepest descent algorithm *all* solutions in the neighbourhood are evaluated, with the best of these then being chosen as the next incumbent. A pseudocode description of this process is given in Figure 3.6.

One advantage of using steepest descent over random descent is that it is abundantly clear when a local optimum has been reached (the algorithm will not be able to identify any solution $S' \in N(S)$ that is better than $S$). Tabu search extends steepest descent by offering a mechanism for escaping these local optima. It does this by

| STEEPEST-DESCENT (*improved* ← **true**) |
| --- |
| (1)  Produce an initial solution $S$ |
| (2)  **while** (*improved* = **true**) **do** |
| (3)       Choose $S' \in N(S) : \forall S'' \in N(S),\ f(S') \leq f(S'')$ |
| (4)       **if** $f(S') < f(S)$ **then** |
| (5)            $S \leftarrow S'$ |
| (6)       **else** |
| (7)            *improved* ← **false** |

**Fig. 3.6** The Steepest Descent algorithm. Here we are seeking to minimise the objective function $f$

also allowing worsening moves to be made when they are seen the be the best available in the current neighbourhood. To avoid cycling, tabu search also then makes use of a memory structure called a tabu list which keeps track of previously visited solutions, and bans the algorithm from returning to these for a certain period of time. This therefore encourages the algorithm to enter new parts of the solution space.

As we have discussed, the papers of Chams et al. (1987) and Hertz and de Werra (1987) suggested some time ago that both the simulated annealing and tabu search metaheuristics are suitable for helping to tackle graph colouring problems. A tabu search method called TABUCOL in particular has proved to be very popular, both when used in isolation and when used as an improvement procedure as part of broader algorithmic schemes. This algorithm will be discussed further in Chapter 4.

In more recent years, many methods for exploring the space of complete improper $k$-colourings have also been proposed, including techniques based on:

- Evolutionary algorithms (Dorne and Hao, 1998; Eiben et al., 1998; Fleurent and Ferland, 1996; Galinier and Hao, 1999);
- Iterated local search (Chiarandini and Stützle, 2002; Paquete and Stützle, 2002);
- GRASP algorithms (Laguna and Marti, 2001);
- Variable neighbourhood search (Avanthay et al., 2003);
- Ant colony optimisation (Thompson and Dowsland, 2008).

Two of the most notable examples from the above list, particularly due to the quality of results that they are reported to produce, are the hybrid evolutionary algorithm of Galinier and Hao (1999) and the ant colony optimisation algorithm of Thompson and Dowsland (2008). Both of these algorithms make use of population-based methods combined with the TABUCOL algorithm. The idea behind this hybridisation is to use the population-based elements of the algorithms to guide the search over the long term, gently directing it towards favourable regions of the solution space, with the TABUCOL element then being used to identify high-quality solutions within these regions. Both of these algorithms will be considered in further detail in Chapter 4.

### 3.2.3 Spaces of Partial, Proper k-Colourings

A further strategy for graph colouring that has perhaps received less attention historically involves exploring the space of proper *partial* solutions. In general, this scheme again involves stipulating a fixed number of colours $k$ at the outset; however, when vertices are encountered that cannot be feasibly assigned to a colour, they are transferred to a set of uncoloured vertices $U$. The aim is to then make changes to the solution so that all vertices in $U$ can be feasibly coloured, resulting in $U = \emptyset$. If this goal is achieved, $k$ can then be reduced and the algorithm repeated, as with the previous scheme.

An effective example of this strategy is the PARTIALCOL algorithm of Blöchliger and Zufferey (2008). This approach uses tabu search and operates in a very similar fashion to the TABUCOL algorithm, albeit with a different neighbourhood operator. Specifically, a move in the solution space is achieved as follows:

- Select an uncoloured vertex $v \in U$ and assign it to a colour class $S_j$ in the partial solution $S$. Next, take all vertices $u \in S_j$ that are adjacent to $v$ and move them from $S_j$ to $U$.

In their work, Blöchliger and Zufferey (2008) make use of the simple objective function $f_3 = |U|$ to evaluate solutions. A second objective function, $f_4 = \sum_{v \in U} \deg(v)$, is also suggested, but was found to only give better solutions in a small number of cases. This algorithm is discussed in more detail in Chapter 4.

An earlier algorithm using this scheme was also suggested by Morgenstern and Shapiro (1990). This used the above neighbourhood operator and objective function in conjunction with simulated annealing. However, it also employed an additional operator that was periodically applied to the partial solution to help reinvigorate the search process. Specifically, this mechanism shuffled vertices between colour classes in the partial solution while not introducing any clashes. This has the effect of moving the algorithm into a different part of the solution space, without changing its objective function value.

High-quality results based on exploring the space of partial proper $k$-colourings have also been reported by Malaguti et al. (2008). This algorithm is similar to the hybrid evolutionary algorithm of Galinier and Hao (1999) and uses an analogous recombination operator together with a local search procedure based on PARTIALCOL. Their approach also makes use of the objective function $f_4$ in an attempt to sustain evolutionary pressure in the population. Due to its use of proper partial solutions, note that all colour classes built by this algorithm are also independent sets. One novel feature of this work is that, during a run of the EA, a set of these independent sets is maintained and added to. Upon termination of the EA, this set is then used in conjunction with an IP model similar to our final model in Section 3.1.2 in order to try to make further improvements.

### 3.2.4 Combining Solution Spaces

Interesting work has also been carried out by Hertz et al. (2008), who propose
a method for operating in *different* solution spaces during different stages of a
run. Specifically, TABUCOL is used to explore the space of complete improper $k$-
colourings, and PARTIALCOL is used for the space of partial, proper solutions. The
main idea here is that a local optimum in one solution space is not necessarily a
local optimum in another. Hence, when the search is deemed to have stagnated in
one space, a procedure is used to alter the incumbent solution so that it becomes a
member of another space. (For example, a complete improper solution formed by
TABUCOL is converted into a partial proper solution by considering clashing vertices in a random order, and moving them into the set $U$ until no clashes remain.)
The search can then be continued in this new space where further improvements
might be made, with the process being repeated as long as necessary. The authors
also propose a third solution space based on the idea of assigning orientations to
edges in the graph and then trying to minimise the length of the longest paths within
the resultant directed graph (see also the work of Gendron et al. (2007)). The authors note, however, that improvements are rarely achieved during exploration of
the latter space, but that its inclusion can still be useful because it tends to make
large alterations to a solution, helping to diversify the search.

### 3.2.5 Problems Related to Graph Colouring

Concluding this review, it is relevant to note that many of the schemes mentioned
above are also commonly used in algorithms tackling problems related to graph
colouring. For example, we can observe the existence of timetabling algorithms that
use constructive heuristics with backtracking (Carter et al., 1996); algorithms that
allow additional timeslots (colours) in a timetable and then only deal with feasible
solutions (Burke et al., 1995; Cote et al., 2005; Erben, 2001; Lewis and Paechter,
2007); methods that fix the number of timeslots in advance and then allow constraints to be violated (i.e., clashes to occur) (Carrasco and Pato, 2001; Colorni et al.,
1997; Di Gaspero and Schaerf, 2002); and also algorithms that deal with partial
timetables, never allowing constraint violations to occur (Burke and Newall, 1999;
Paechter et al., 1998; Lewis and Thompson, 2015). Similar examples can also be
noted in other related problems such as the frequency assignment problem (Aardel
et al., 2002; Valenzuela, 2001).

## 3.3 Reducing Problem Size

When applying any of the algorithmic techniques mentioned above, it will often be
in our interest to reduce the size of the problem instance at hand by eliminating

vertices and edges. In turn this could lead to shorter run times and/or more accurate results. The following two subsections now discuss ways in which this can be achieved.

### 3.3.1 Removing Vertices and Splitting Graphs

Given a graph $G = (V, E)$:

- Let $u, v \in V$ such that $\Gamma(u) \subseteq \Gamma(v)$. This implies that $u$ and $v$ are nonadjacent. Now let $G' = G - \{u\}$. We can now colour the smaller graph $G'$ using any algorithm. Once a feasible colouring for $G'$ has been established, $u$ can then be reinserted into the graph and assigned to the same colour as vertex $v$.
- Let $C \subseteq V$ be a subgraph of $G$ such that (a) $C$ is a clique and (b) the vertices of $C$ are a separating set (see Definition 2.11). Now label as $G_1, \ldots, G_l$ the components that are formed by deleting $C$ from $G$, and let $G'_1 = G_1 \cup C$, $G'_2 = G_2 \cup C, \ldots, G'_l = G_l \cup C$. Feasible colourings of the smaller subgraphs $G'_1, \ldots, G'_l$ can now be produced separately and then merged into a feasible colouring of $G$.

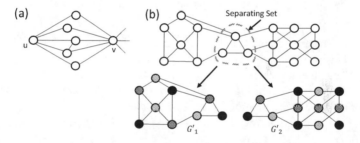

**Fig. 3.7** Examples of graphs that can be reduced in size before colouring

As an example of the first bullet above, consider Figure 3.7(a) where, as required, $\Gamma(u) \subseteq \Gamma(v)$. It is clear from this figure that $u$ can always be assigned to the same colour as $v$. Hence, $u$ can be removed from the graph together with all its incident edges. These can be reinstated once the remaining vertices have been coloured.

To illustrate the second bullet, consider Figure 3.7(b). As indicated, this graph contains a separating set of size 3 which is also a clique. In this case, the two smaller subgraphs $G'_1$ and $G'_2$ can now be coloured separately. If the vertices in the separating set are not allocated to the same colours in each subgraph (as is the case here), then a colour relabelling can be applied to make this so. The subgraphs can then be merged to form a complete feasible colouring for $G$. Note that, by definition, this feature includes cases where a graph $G$ is disconnected (giving $|C| = 0$) or $G$ contains a cut vertex ($|C| = 1$). Also note that $\chi(G) = \max\{\chi(G'_1), \ldots, \chi(G'_l)\}$.

In practice, it is easy to check whether there exist vertices $u, v$ such that $\Gamma(u) \subseteq \Gamma(v)$ and, depending on the topology of the graph, it might be possible to remove many vertices before applying a graph colouring algorithm. The problem of identifying separating sets is also readily solvable by various polynomial-time algorithms (such as the approach of Kanevsky (1993)), and it only takes the addition of a simple checking step to determine whether these separating sets also constitute cliques or not.

In addition to these steps, in situations where we are trying to solve the decision variant of the graph colouring problem (given an integer $k$, identify whether a feasible $k$-colouring exists), we are also able to eliminate all vertices with degrees of less than $k$. That is, we can reduce the size of $G$ by removing all vertices from the set $\{v \in V : \deg(v) < k\}$. This is permitted since, obviously, vertices with fewer than $k$ adjacent vertices will always have a feasible colour from the set $\{1, \ldots, k\}$ to which they can be assigned. Hence these colours can be allocated to these vertices once the remaining subgraph has been coloured.

### 3.3.2 Extracting Independent Sets

A further method for reducing the size of a graph involves the identification and removal of independent sets. A suitable process can be summarised as follows. Given a graph $G = (V, E)$:

1. Let $G' = G$. Identify an independent set $I_1$ in $G'$ and remove it. Repeat this step on $G'$ a further $l - 1$ times to form a set of $l$ disjoint independent sets $\{I_1, I_2, \ldots, I_l\}$. Call $G'$ the residual graph.
2. Next, use any graph colouring algorithm to find a feasible colouring for the residual graph $G'$. Call this solution $\mathcal{S}' = \{S_1, \ldots, S_k\}$.
3. A feasible $(k + l)$-colouring for the original graph $G$ is obtained by setting $\mathcal{S} = \mathcal{S}' \cup \{I_1, I_2, \ldots, I_l\}$.

For Step 1 above it is usually helpful to identify large independent sets because this will leave us with a smaller residual graph. Recall, however, that the problem of identifying the maximum independent set in a graph is itself an NP-hard problem, implying the need for heuristics and/or approximation algorithms in most cases. Methods for identifying large independent sets in a graph range from simple greedy techniques such as the RLF algorithm (Section 2.4) and the heuristics of Chams et al. (1987) to advanced metaheuristic algorithms such as the tabu search approach of Wu and Hao (2012). A typical local search-based scheme might operate by seeking to establish an independent set of size $q$ in a graph $G = (V, E)$. Its solution space might be the set of all $q$-subsets of $V$ (that is, a candidate solution $I \subseteq V$ with $|I| = q$), and its objective function, $f_5$, might simply be a count on the number of edges in the subgraph of $G$ induced by $I$:

$$f_5(I) = \sum_{\forall u, v \in I} g(u, v) \tag{3.16}$$

where

$$g(u,v) = \begin{cases} 1 \text{ if } \{u,v\} \in E \\ 0 \text{ otherwise.} \end{cases}$$

For this solution space, a suitable neighbourhood operator might simply swap a vertex $u \in I$ with a vertex $v \in (V - I)$, with the aim of identifying a solution $I$ with zero cost, leaving us with an independent set of size $q$. If this set is deemed large enough, it might then be removed from the graph; else $q$ can be increased and the process repeated. We might also seek to fulfil additional criteria, such as identifying an independent set that, when removed, leaves a residual graph with the fewest number of edges, thereby hopefully giving it a lower chromatic number.

Note that if we choose to reduce a problem's size by extracting independent sets, a suitable balance will need to be struck between the time dedicated to this task and the time used for colouring the residual graph itself. We should also be mindful of the fact that extracting the *wrong* independent sets may also prevent us from being able to identify the optimal solution to the original graph colouring problem.

Finally, note that the problem of identifying a maximum independent set in a graph $G$ is actually equivalent to identifying a maximum clique in $G$'s complement graph $\bar{G} = (V, \bar{E})$ (where $\bar{E} = \{\{u,v\} : \{u,v\} \notin E\}$). A helpful survey on heuristic algorithms for both of these problems is provided by Pelillo (2009).

# Chapter 4
# Algorithm Case Studies

In this chapter we present detailed descriptions of six high-performance algorithms for the graph colouring problem. Implementations of each of these can be found in the online suite of graph colouring algorithms described in Section 1.6.1 and Appendix A.1. In Section 4.2 onwards we then compare the performance of these algorithms over a wide range of graphs in order to gauge their relative strengths and weaknesses.

## 4.1 Algorithm Descriptions

### 4.1.1 The TABUCOL Algorithm

As we mentioned in the previous chapter, since its proposal by Hertz and de Werra in 1987, TABUCOL has been used as a local search subroutine in a number of high-performing hybrid algorithms, including those of Avanthay et al. (2003), Dorne and Hao (1998), Galinier and Hao (1999), and Thompson and Dowsland (2008). The specific version of TABUCOL that we consider here is the so-called "improved" variant, which was originally used by Galinier and Hao (1999). The various features of this algorithm are now reviewed.

TABUCOL operates in the space of complete improper $k$-colourings using an objective function that simply counts the number of clashes, as defined by $f_2$ in Equation (3.15). Given a candidate solution $\mathcal{S} = \{S_1, \ldots, S_k\}$, moves in the solution space are performed by selecting a vertex $v \in S_i$ whose assignment to colour class $S_i$ is currently causing a clash, and then assigning it to a new colour class $S_j \neq S_i$. Note that previous incarnations of this algorithm also allowed nonclashing vertices to be moved between colours, though this is generally seen to worsen performance (Galinier and Hertz, 2006).

The tabu list of the algorithm is stored using a matrix $\mathbf{T}_{n \times k}$. If, at iteration $l$ of the algorithm, the neighbourhood operator transfers a vertex $v$ from $S_i$ to $S_j$, then the

© Springer International Publishing Switzerland 2016
R.M.R. Lewis, *A Guide to Graph Colouring*,
DOI 10.1007/978-3-319-25730-3_4

element $T_{vi}$ is set to $l + t$, where $t$ is a positive integer that will be defined presently. This signifies that the moving of $v$ back to colour class $S_i$ is *tabu* (i.e., disallowed) for $t$ iterations of the algorithm (or, in other words, that $v$ cannot be moved back to $S_i$ until at least iteration $l + t$). Note that this has the effect of making *all* solutions containing the assignment of vertex $v$ to $S_i$ tabu for $t$ iterations.

As is typical in applications of tabu search, in each iteration of TABUCOL the entire set of neighbouring solutions is considered. That is, the cost of moving each clashing vertex into all other $k - 1$ colour classes is evaluated. This process consumes the majority of the algorithm's execution time; however, it can be sped up considerably through the use of appropriate data structures. To explain, let $x$ denote the number of vertices involved in a clash in the current solution $S$. This leads to $x(k-1)$ members in the set of neighbouring solutions $N(S)$. (Obviously, there is a strong positive correlation between $x$ and the objective function, so better solutions will tend to have smaller neighbourhoods.) A naïve implementation of the TABU-COL would set about separately performing the $x(k-1)$ different neighbourhood moves and evaluating all the resulting solutions. However, this is not necessary, particularly because only two colour classes are effected by each neighbourhood move.

A more efficient approach involves making use of an additional matrix $\mathbf{C}_{n \times k}$ where, given the current solution $S = \{S_1, \ldots, S_k\}$, element $C_{vj}$ denotes the number of vertices in colour class $S_j$ that are adjacent to vertex $v$. When an initial solution is generated, all elements in $\mathbf{C}$ will need to be calculated. However, in each subsequent iteration of TABUCOL, the act of moving a vertex $v$ from $S_i$ to $S_j$ will result in a new solution $S'$ whose cost is simply:

$$f_2(S') = f_2(S) + C_{vj} - C_{vi}. \tag{4.1}$$

Since $f_2(S)$ will already be known, this means that the cost of all neighbouring solutions can be determined by simply scanning each row of $\mathbf{C}$ corresponding to clashing vertices in $S$.

Once a move has been selected and performed (i.e., once $v$ has been moved from $S_i$ to $S_j$), the matrix $\mathbf{C}$ can be updated using the procedure shown in Figure 4.1. As shown in this pseudocode, neighbours of $v$ are now marked as being adjacent to one fewer vertex in colour class $S_i$ and one additional vertex in colour class $S_j$.

---

| UPDATE-C $(v, i, j)$ |
| --- |
| (1) **forall** $u \in \Gamma(v)$ **do** |
| (2)      $C_{ui} \leftarrow C_{ui} - 1$ |
| (3)      $C_{uj} \leftarrow C_{uj} + 1$ |

---

**Fig. 4.1** Procedure for updating the matrix $\mathbf{C}$ once TABUCOL has moved a vertex $v$ from colour $i$ to colour $j$. As usual, $\Gamma(v)$ denotes the set of all vertices adjacent to vertex $v$

Having evaluated all neighbouring solutions, TABUCOL selects and performs the non-tabu move that brings about the largest decrease (or failing that, the smallest increase) in cost. Any ties in this criterion are broken randomly. In addition, TABU-COL also employs an *aspiration criterion* which allows tabu moves to be made on occasion. Specifically, they are permitted if they are seen to improve on the best solution found so far during the run. This is particularly helpful if a tabu move is seen to lead to a solution $S'$ with zero cost, at which point the algorithm can halt. Finally, if *all* moves are seen to be tabu, then a vertex $v \in V$ is selected at random and moved to a new randomly selected colour class. The tabu list is then updated as usual.

In the version of TABUCOL that we use here, an initial candidate solution is constructed by taking a random ordering of the vertices and applying a modified version of the GREEDY algorithm in which only $k$ colours are permitted. Thus, if vertices are encountered that cannot be assigned to any of the $k$ colours without inducing a clash, these are assigned to one of the existing colours randomly. Of course, we could use more sophisticated constructive methods here, but it is stated by both Galinier and Hertz (2006) and Blöchliger and Zufferey (2008) that the method of initial solution generation is not critical in TABUCOL's performance.

Finally, with regard to the *tabu tenure*, Galinier and Hao (1999) have suggested making $t$ a random variable that is proportional to the incumbent solution's cost. The idea here is that when the incumbent solution is poor, its high cost will lead to large values for $t$, which will hopefully force the algorithm into different regions of the solution space where better solutions can be found. On the other hand when the incumbent solution has a low cost, the algorithm should focus on the current region by using low values for $t$. Galinier and Hao (1999) suggest using $t = 0.6f_2 + r$, where $r$ is an integer uniformly selected from the range 0 to 9 inclusive. These particular settings have been used in various other applications of TABUCOL (Blöchliger and Zufferey, 2008; Galinier and Hao, 1999; Thompson and Dowsland, 2008) and are generally thought to give good results; however, it should be noted that other schemes for determining $t$ are likely to be more appropriate for certain graphs.

### 4.1.2 The PARTIALCOL Algorithm

The PARTIALCOL algorithm of Blöchliger and Zufferey (2008) operates in a similar fashion to TABUCOL in that it uses the tabu search metaheuristic to seek a proper $k$-colouring. However, in contrast to TABUCOL, PARTIALCOL does not consider improper solutions; instead, vertices that cannot be assigned to any of the $k$ colours without causing a clash are put into a set of uncoloured vertices $U$. The aim of PARTIALCOL is to thus make alterations to the partial solution $S$ so that $U$ can be emptied, giving $f_3 = |U| = 0$ and, consequently, a feasible $k$-coloured solution.

Because of its use of partial proper solutions, the neighbourhood operator of PARTIALCOL is somewhat different from that of TABUCOL. Specifically, a move in the solution space is achieved by selecting an uncoloured vertex $v \in U$ and assigning it to a colour class $S_j \in S$. The move is then completed by taking all vertices $u \in S_j$ that are adjacent to $v$ and transferring them from $S_j$ into $U$. Having performed such a move, all corresponding elements $T_{uj}$ in the tabu list are then marked as tabu for the next $t$ iterations of the algorithm.

In each iteration of PARTIALCOL, the complete set of $|U| \times k$ neighbouring solutions is examined. The move to be performed is then chosen using the same criteria as TABUCOL. As with TABUCOL the matrix $\mathbf{C}$ can again be used to speed up the process of evaluating the neighbourhood set. In this case, the act of moving vertex $v$ from $U$ to colour class $S_j$ leads to a new solution $\mathcal{S}'$ whose cost is simply:

$$f_3(\mathcal{S}') = f_3(\mathcal{S}) + C_{vj} - 1. \tag{4.2}$$

Once a move has been performed (that is, the vertex $v \in U$ has been transferred to $S_j$ and all vertices in the set $\{u \in S_j : u \in \Gamma(v)\}$ have been moved to $U$), the $\mathbf{C}$ matrix is updated using the procedure given in Figure 4.2.

| UPDATE-C $(v, j)$ |
| --- |
| (1) **forall** $u \in \Gamma(v)$ **do** |
| (2) $\qquad C_{uj} \leftarrow C_{uj} + 1$ |
| (3) $\qquad$ **if** $c(u) = j$ **then** |
| (4) $\qquad\qquad$ **forall** $w \in \Gamma(u)$ **do** |
| (5) $\qquad\qquad\qquad C_{wj} \leftarrow C_{wj} - 1$ |

**Fig. 4.2** Procedure for updating $\mathbf{C}$ once PARTIALCOL has moved vertex $v$ from the set $U$ to colour $j$

An initial solution to PARTIALCOL is generated using a greedy process analogous to that of TABUCOL. The only difference is that when vertices are encountered for which there exists no clash-free colours, these are put into the set $U$. The only other operational difference between the two algorithms relates to the calculation of the tabu tenure $t$. In their original paper, Blöchliger and Zufferey (2008) use an algorithm variant known as FOO-PARTIALCOL. Here, FOO abbreviates "Fluctuation Of the Objective-function", and indicates their use of a mechanism that alters $t$ based on the algorithm's search progress. In essence, if during a run the objective function has not altered for a lengthy period of time, it is assumed that the search has stagnated in a particular region of the solution space and so $t$ is increased to try to encourage the algorithm to leave this region. Similarly, when the objective function is seen to be fluctuating, $t$ is slowly reduced, counteracting these effects. Note that this scheme requires values to be assigned to a number of parameters, the meanings of which are described by Blöchliger and Zufferey (2008). In our case, we choose to use settings recommended by the authors and these are included in our source code of this algorithm. We are perfectly at liberty to use other simpler schemes for calculating $t$ if required, however.

## 4.1.3 The Hybrid Evolutionary Algorithm (HEA)

The third algorithm that we shall consider is the hybrid evolutionary algorithm (HEA) of Galinier and Hao (1999). The HEA operates by maintaining a population of candidate solutions that are evolved via a problem-specific recombination operator and a local search method. Like TABUCOL, the HEA operates in the space of complete improper $k$-colourings using cost function $f_2$.

The algorithm begins by creating an initial population of candidate solutions. Each member of this population is formed using a modified version of the DSATUR algorithm for which the number of colours $k$ is fixed at the outset. To provide diversity between members, the first vertex is selected at random and assigned to the first colour. The remaining vertices are then taken in sequence according to the maximum saturation degree (with ties being broken randomly) and assigned to the lowest indexed colour class $S_i$ seen to be feasible (where $1 \leq i \leq k$). When vertices are encountered for which no feasible colour class exists, these are kept to one side and are assigned to random colour classes at the end of this process. Upon construction of this initial population, an attempt is then made to improve each member by applying the local search routine.

As is typical for an evolutionary algorithm, for the remainder of the run the algorithm evolves the population using recombination, mutation, and evolutionary pressure. In each iteration two parent solutions $S_1$ and $S_2$ are selected from the population at random, and copies of these are used in conjunction with the recombination operator to produce one child solution $S'$. This child is then improved via the local search operator, and is inserted into the population by replacing the weaker of its two parents. Note that there is no bias towards selecting fitter parents for recombination; rather evolutionary pressure only exists due to the offspring replacing their weaker parent (regardless of whether the parent has a better cost than its child).

| | Parent $S_1$ | Parent $S_2$ | Offspring $S'$ | Comments |
|---|---|---|---|---|
| 1) | $\{\{v_1, v_2, v_3\}, \{v_4, v_5, v_6, v_7\}, \{v_8, v_9, v_{10}\}\}$ | $\{\{v_3, v_4, v_5, v_7\}, \{v_1, v_6, v_9\}, \{v_2, v_8, v_{10}\}\}$ | $\{\}$ | To start, the offspring solution $S = \emptyset$. |
| 2) | $\{\{v_1, v_2, v_3\}, \{v_8, v_9, v_{10}\}\}$ | $\{\{v_3\}, \{v_1, v_9\}, \{v_2, v_8, v_{10}\}\}$ | $\{\{v_4, v_5, v_6, v_7\}\}$ | Select the colour class with most vertices and copy it into $S'$. (Class $\{v_4, v_5, v_6, v_7\}$ from $S_1$ in this case.) Delete the copied vertices from both $S_1$ and $S_2$. |
| 3) | $\{\{v_1, v_3\}, \{v_9\}\}$ | $\{\{v_3\}, \{v_1, v_9\}\}$ | $\{\{v_4, v_5, v_6, v_7\}, \{v_2, v_8, v_{10}\}\}$ | Select the largest colour class in $S_2$ and copy it into $S'$. Delete the copied vertices from both $S_1$ and $S_2$. |
| 4) | $\{\{v_9\}\}$ | $\{\{v_9\}\}$ | $\{\{v_4, v_5, v_6, v_7\}, \{v_2, v_8, v_{10}\}, \{v_1, v_3\}\}$ | Select the largest colour class in $S_1$ and copy it into $S'$. Delete the copied vertices from both $S_1$ and $S_2$. |
| 5) | $\{\{v_9\}\}$ | $\{\{v_9\}\}$ | $\{\{v_4, v_5, v_6, v_7\}, \{v_2, v_8, v_{10}, v_9\}, \{v_1, v_3\}\}$ | Having formed $k$ colour classes, assign any missing vertices to random colours to form a complete but not necessarily proper offspring solution $\mathcal{S}$. |

**Fig. 4.3** Example application of the Greedy Partition Crossover of Galinier and Hao (1999), using $k = 3$

The recombination operator proposed Galinier and Hao (1999) is the so-called Greedy Partition Crossover (GPX). The idea behind GPX is to construct offspring using large colour classes inherited from both parent solutions. A demonstration of how this is done is given in Figure 4.3. As shown, the largest (not necessarily proper) colour class from the parents is first selected and copied into the offspring (ties are broken randomly). In order to avoid duplicate vertices occurring in the offspring at a later stage, these copied vertices are then removed from both parents. To form the next colour, the other (modified) parent is then considered and, again, the largest colour class is selected and copied into the offspring, before these vertices are removed from both parents. This process is continued by alternating between the parents until the offspring's $k$ colour classes have been formed.

At this point, each colour class in the offspring will be a subset of a colour class existing in one or both of the parents. That is:

$$\forall S_i \in \mathcal{S}' \; \exists S_j \in (\mathcal{S}_1 \cup \mathcal{S}_2) \; : \; S_i \subseteq S_j \qquad (4.3)$$

where $\mathcal{S}'$, $\mathcal{S}_1$, and $\mathcal{S}_2$ represent the offspring, and the first and second parents respectively. However, some vertices may be missing in the offspring (as is the case with vertex $v_9$ in Figure 4.3). This issue is resolved by assigning the missing vertices to random colour classes.

Once a complete offspring solution is formed, it is then modified and improved via a local search procedure before being inserted into the population. For this purpose the TABUCOL algorithm is used for a fixed number of iterations $I$ using the same tabu tenure scheme as described in Section 4.1.1. In their original paper, Galinier and Hao (1999) present results for a small sample of problem instances and manually tune $I$ for each case. In our case we choose not to follow this strategy and require a setting for $I$ to be determined automatically by the algorithm. We also need to be wary that if $I$ is set too low, then insufficient local search will be carried out on each newly created solution, while an $I$ that is too high will result in too much effort being placed on local search as opposed to the global search carried out by the evolutionary operators. Ultimately we choose to settle on $I = 16n$, which corresponds roughly to the settings used in the most successful runs reported by Galinier and Hao (1999). In all cases, we also use a population size of 10, as recommended by the authors.

### 4.1.4 The ANTCOL Algorithm

Like the HEA, the ANTCOL algorithm of Thompson and Dowsland (2008) is another metaheuristic-based method that combines global and local search operators, in this case using the ant colony optimisation (ACO) metaheuristic.

ACO is an algorithmic framework that was originally inspired by the way in which real ants determine efficient paths between food sources and their colonies. In their natural habitat, when no food source has been identified, ants tend to wander about randomly. However, when a food source is found, the discovering ants will take some of this back to the colony leaving a pheromone trail in their wake. When other ants discover this pheromone, they are less likely to continue wandering at random, but may instead follow the trail. If they go on to discover the same food source, they will then follow the pheromone trail back to the nest, adding their own pheromone in the process. This encourages further ants to follow the trail. In addition to this, pheromones on a trail also tend to evaporate over time, reducing the chances of an ant following it. The longer it takes for an ant to traverse a path, the more time the pheromones have to evaporate; hence shorter paths tend to see a more rapid build-up of pheromone, making other ants more likely to follow it and deposit their own pheromone. This positive feedback eventually leads to all ants following a single, efficient path between the colony and food source.

As might be expected, initial applications of ACO were aimed towards problems such as the travelling salesman problem and vehicle routing problems, where we seek to identify efficient paths for visiting the vertices of a graph (see for example the work of Dorigo et al. (1996) and Rizzoli et al. (2007)). However, applications to many other problems have also been made.

The idea behind the ANTCOL algorithm is to use ants to produce individual candidate solutions. During a run each ant produces its solution in a nondeterministic manner, using probabilities based on heuristics and also on the quality of solutions produced by previous ants. In particular, if previous ants have identified features that are seen to lead to better-than-average solutions, the current ant is more likely to include these features in its own solution, generally leading to a reduction in the number of colours during the course of a run.

A full description of the ANTCOL algorithm is provided in Figure 4.4. As shown in the pseudocode, in each cycle of the algorithm (lines (3) to (19)), a number of ants each produce a complete, though not necessarily feasible, solution. In line (16) the details of each of these solutions are then added to a trail update matrix $\delta$ and, at the end of a cycle, the contents of $\delta$ are used together with an evaporation rate $\rho$ to update the global trail matrix $t$.

At the start of each cycle, each individual ant attempts to construct a solution using the procedure BUILDSOLUTION. This is based on the RLF method (see Section 2.4) which, we recall, operates by building up each colour class in a solution one at a time. Also recall that during the construction of each class $S_i \in S$, RLF makes use of two sets: $X$, which contains uncoloured vertices that can currently be added to $S_i$ without causing a clash; and $Y$, which holds the uncoloured vertices that *cannot* be feasibly added to $S_i$. The modifications to RLF that BUILDSOLUTION employs are as follows:

- In the procedure a maximum of $k$ colour classes is permitted. Once these have been constructed, any remaining vertices are left uncoloured.
- The first vertex to be assigned to each colour class $S_i$ ($1 \leq i \leq k$) is chosen randomly from the set $X$.
- In remaining cases, each vertex $v$ is then assigned to colour $S_i$ with probability

---

ANTCOL $(G = (V,E))$

---

(1)  $t_{uv} \leftarrow 1 \ \forall u,v \in V : u \neq v$
(2)  $k = n$
(3)  **while** (**not** stopping condition) **do**
(4)  $\quad \delta_{uv} \leftarrow 0 \ \forall u,v \in V : u \neq v$
(5)  $\quad best \leftarrow k$
(6)  $\quad foundFeasible \leftarrow$ **false**
(7)  $\quad$ **for** $(ant \leftarrow 1 \ \textbf{to} \ nants)$ **do**
(8)  $\quad\quad S \leftarrow$ BUILDSOLUTION$(k)$
(9)  $\quad\quad$ **if** $(S$ is a partial solution) **then**
(10) $\quad\quad\quad$ Randomly assign uncoloured vertices to colour classes in $S$
(11) $\quad\quad\quad$ Run TABUCOL
(12) $\quad\quad$ **if** $(S$ is feasible) **then**
(13) $\quad\quad\quad foundFeasible \leftarrow$ **true**
(14) $\quad\quad\quad$ **if** $(|S| \leq best)$ **then**
(15) $\quad\quad\quad\quad best \leftarrow |S|$
(16) $\quad\quad\quad\quad \delta_{uv} \leftarrow \delta_{uv} + F(S) \ \forall u,v : c(u) = c(v) \wedge u \neq v$
(17) $\quad t_{uv} \leftarrow \rho \times t_{uv} + \delta_{uv} \ \forall u,v \in V : u \neq v$
(18) $\quad$ **if** $(foundFeasible=$**true**$)$ **then**
(19) $\quad\quad k \leftarrow best-1$

---

**Fig. 4.4** The ANTCOL algorithm. At termination, the best feasible solution found uses $k + 1$ colours

$$
P_{vi} = \begin{cases} \dfrac{\tau_{vi}^{\alpha} \times \eta_{vi}^{\beta}}{\sum_{u \in X}(\tau_{ui}^{\alpha} \times \eta_{ui}^{\beta})} & \text{if } v \in X \\ 0 & \text{otherwise} \end{cases} \tag{4.4}
$$

where $\tau_{vi}$ is calculated

$$
\tau_{vi} = \frac{\sum_{u \in S_i} t_{uv}}{|S_i|}. \tag{4.5}
$$

Note that the calculation of $\tau_{vi}$ makes use of the global trail matrix $t$, meaning that higher values are associated with combinations of vertices that have been assigned the same colour in previous solutions. The value $\eta_{vi}$, meanwhile, is associated with a heuristic rule which, in this case, is the degree of vertex $v$ in the graph induced by the set of currently uncoloured vertices $X \cup Y$. Larger values for $\tau_{vi}$ and $\eta_{vi}$ thus contribute to larger values for $P_{vi}$, encouraging vertex $v$ to be assigned to colour class $S_i$. The parameters $\alpha$ and $\beta$ are used to control the relative strengths of $\tau$ and $\eta$ in the equation.

The ANTCOL algorithm also makes use of a "multi-sets" operator in the BUILD-SOLUTION procedure. Since the process of constructing a colour class is probabilistic, the operator makes $\nu$ separate attempts to construct each colour class. It then selects the one that results in the minimum number of edges in the graph induced by the set of remaining uncoloured vertices $Y$ (since such graphs will tend to feature lower chromatic numbers).

On completion of BUILDSOLUTION, the generated solution $S$ will be proper, but could be partial. If the latter is true, all uncoloured vertices are assigned to random colour classes to form a complete, improper solution, and TABUCOL is run for $I$ iterations. Details on the solution are then written to the trail update matrix $\delta$ using the evaluation function:

$$F(S) = \begin{cases} 1/f_2 & \text{if } f_2 > 0 \\ 3 & \text{otherwise.} \end{cases} \tag{4.6}$$

This means that higher-quality solutions contribute larger values to $\delta$, encouraging their features to be included in solutions produced by future ants.

The parameters used in our application, and recommended by Thompson and Dowsland (2008), are as follows: $\alpha = 2$, $\beta = 3$, $\rho = 0.75$, *nants* $= 10$, $I = 2n$, and $v = 5$. The tabu tenure scheme of TABUCOL is the same as in previous descriptions.

## 4.1.5 The Hill-Climbing (HC) Algorithm

In contrast to the preceding four algorithms, the Hill-Climbing (HC) algorithm of Lewis (2009) operates in the space of feasible solutions, with the initial solution being formed using the DSATUR heuristic. During a run, the algorithm operates on a single feasible solution $S = \{S_1, \ldots S_{|S|}\}$ with the aim of minimising $|S|$. To begin, a small number of colour classes are removed from $S$ and are placed into a second set $T$, giving two partial proper solutions. A specialised local search procedure is then run for $I$ iterations. This attempts to feasibly transfer vertices from colour classes in $T$ into colour classes in $S$ such that both $S$ and $T$ remain proper. If successful, this has the effect of increasing the cardinality of the colour classes in $S$ and may also empty some of the colour classes in $T$, reducing the total number of colours being used. At the end of the local search procedure, all colour classes in $T$ are copied back into $S$ to form a feasible solution.

The first iteration of the local search procedure operates by considering each vertex $v$ in $T$ and checking whether it can be feasibly transferred into any of the colour classes in $S$. If this is the case, such transfers are performed. The remaining iterations of the procedure then operate as follows. First, an alteration is made to a randomly selected pair of colour classes $S_i, S_j \in S$ using either a Kempe chain interchange or a pair swap (see Definitions (3.1) and (3.2)). Since this will usually alter the make-up of two colour classes,[1] this then raises the possibility that other vertices in $T$ can now also be moved to $S_i$ or $S_j$. Again, these transfers are made

---

[1] Note that in some cases a Kempe chain will contain all vertices in both colour classes: that is, the graph induced by $S_i \cup S_j$ will form a connected bipartite graph. Kempe chains of this type are known as *total*, and interchanging their colours serves no purpose since this only results in the two colour classes being relabelled. Consequently total Kempe chains are ignored by the algorithm.

if they are seen to retain feasibility. The local search procedure continues in this fashion for $I$ iterations.

On completion of the local search procedure, the independent sets in $\mathcal{T}$ are copied back into $\mathcal{S}$ to form a feasible solution. The independent sets in $\mathcal{S}$ are then ordered according to some (possibly random) heuristic, and a new solution $\mathcal{S}'$ is formed by constructing a permutation of the vertices in the same manner as that of the Iterated Greedy algorithm (see Section 3.2.1) and then applying the GREEDY algorithm. This latter operation is intended to generate large alterations to the incumbent solution, which is then passed back to the local search procedure for further optimisation. Note that none of the stages of this algorithm allow the number of colour classes being used to increase, thus providing its hill-climbing characteristics.

As with the previous algorithms, a number of parameters have to be set with this algorithm, each that can influence its performance. The values used in our experiments here were determined in preliminary tests and according to those reported by Lewis (2009). For the local search procedure, independent sets are moved into $\mathcal{T}$ by considering each $S_i \in \mathcal{S}$ in turn and transferring it with probability $1/|\mathcal{S}|$. The local search procedure is then run for $I = 1,000$ iterations, and in each iteration the Kempe chain and swap neighbourhoods are called with probabilities 0.99 and 0.01 respectively. Finally, when constructing the permutation of the vertices for passing to the GREEDY algorithm, the independent sets are ordered using the same 5:5:3 ratio as detailed in Section 3.2.1.

### 4.1.6 The Backtracking DSATUR Algorithm

The sixth and final algorithm considered in this chapter is the backtracking approach of Korman (1979). Essentially, this operates in the same manner as the basic backtracking approach discussed in Section 3.1.1, though with the following modifications:

- The initial order of the vertices is determined by the DSATUR algorithm. Hence vertices with the fewest available colours are coloured first, with ties being broken by the degrees, and further ties being broken randomly.
- After performing a backward step, vertices are dynamically reordered so that the next vertex to be coloured is also the one with the fewest available colours. If the vertex has no feasible colours available, the algorithm takes a further backward step.

An example run-through of this algorithm is shown in Figure 4.5. This should be interpreted in the same manner as Figure 3.1. Note that a number of parameters can be set when applying this algorithm, some of which might alter the performance quite drastically. These include specifying the maximum number of branches that can be considered at each node of the tree and prohibiting branching at certain levels of the tree. In practice, it is not obvious how these settings might be chosen a priori for individual graphs, so in our case we opt for the most natural configuration, which

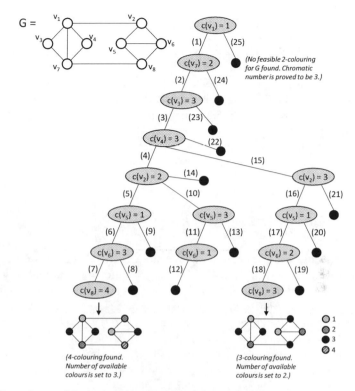

**Fig. 4.5** Example run of the backtracking algorithm of Korman (1979)

is to simply attempt a complete exploration of the search tree.[2] This means that the algorithm is exact under excess time, though of course such run-lengths will not be possible in most cases.

## 4.2  Algorithm Comparison

In this section we now compare the above six algorithms using a selection of different graph types. As with our comparison of constructive algorithms in Chapter 2, we begin by considering random graphs. We then go on to consider a further type of artificially generated graph, the flat graph, before looking more closely at sets of graphs arising in two real-world practical problems, namely university timetabling and social networking.

As with our previous experiments, computational effort for these algorithms is measured by counting the number of constraint checks (see Section 1.6.1). Due to the operational differences of the algorithms, during a run solution quality is mea-

---

[2] These parameters can be altered in the implementation, however.

sured by simply observing the smallest number of colours used in a feasible solution up until that point. Note that because TABUCOL, PARTIALCOL and the HEA operate using infeasible solutions, settings for $k$ are also required which might then need to be modified during a run. In our case initial values are determined by executing DSATUR on each instance and setting $k$ to the number of colours used in the resultant solution. During runs, $k$ is then decremented by 1 each time a feasible $k$-colouring is found, with the algorithms being restarted. In all trials a computation limit of $5 \times 10^{11}$ constraint checks was imposed. This value is chosen to be deliberately high in order to provide some notion of excess time in the trials. Example run times (in seconds) using this computation limit are given in Table 4.6 later.

### 4.2.1 Artificially Generated Graphs

According to Definition 2.15, random graphs are generated such that each pair of vertices is made adjacent with probability $p$. For the following experiments we used values of $p$ ranging from 0.05 (sparse) to 0.95 (dense), incrementing in steps of 0.05, with $n \in \{250, 500, 1000\}$. Twenty-five instances were generated in each case.

The second type of artificial graph we consider are flat graphs. These are produced by taking a graph $G = (V, E = \emptyset)$ and then partitioning the $n$ vertices into $q$ almost equi-sized independent sets (i.e., each set contains either $\lfloor n/q \rfloor$ or $\lceil n/q \rceil$ vertices). Edges are then added between pairs of vertices in different independent sets with probability $p$ in such a way that the variance in vertex degrees is kept to a minimum.

It is well known that $q$-coloured solutions to flat graphs are quite easy to achieve for most values of $p$. This is because for lower values for $p$, problems will be under-constrained, perhaps giving $\chi(G) < q$, and making $q$-coloured solutions easily identifiable. On the other hand, high values for $p$ can result in over-constrained problems with prominent global optima that are easily discovered. Hard-to-solve $q$-colourable graphs are known to occur for a region of $p$'s at the boundary of these extremes, commonly termed the *phase transition region* (Cheeseman et al., 1991; Turner, 1988). Flat graphs, in particular, are known to have rather pronounced phase transition regions because each colour class and vertex degree is deliberately similar, implying a lack of heuristic information for algorithms to exploit.

For our experiments, flat graphs were generated using publicly available software designed by Joseph Culberson which can be downloaded at web.cs.ualberta.ca/~joe/coloring. Graphs were produced for $q \in \{10, 50, 100\}$ using various settings of $p$ in and around the phase transition regions. In each case we used $n = 500$, implying approximately 50, ten, and five vertices per colour respectively. Twenty instances were generated in each case.

Note that according to the structure of random graphs, vertex degrees are characterised by the binomial distribution $B(n - 1, p)$. This means that the standard deviation of the vertex degrees, calculated

$$\sigma = \sqrt{(n-1)p(1-p)}, \tag{4.7}$$

does not exceed 15.8 in this test set of random graphs. It also implies that the degree coefficient of variation (CV), which is defined as the ratio of the standard deviation to the mean $\sigma/\mu$, never exceeds 28% (being maximised at $n = 250$, $p = 0.05$). In a similar fashion, flat graphs are constructed such that variance in degrees is minimised, and for our generated instances this means that the CV never exceeds 28.5%. Compared to many of the more real-world graphs considered later, these values imply a rather high level of vertex homogeneity (i.e., vertices tend to "look the same"), helping to explain some of the following results.

### 4.2.1.1 Performance on Random Graphs

Table 4.1 shows the number of colours used in solutions produced by the six algorithms for random graphs with edge probability $p = 0.5$ and varying numbers of vertices. The results indicate that for the smaller graphs ($n = 250$), the TABUCOL, PARTIALCOL and HEA algorithms produce solutions with fewer colours than the remaining algorithms.[3] However, no statistical difference between these three algorithms is apparent. For larger graphs however, the HEA produces the best results, allowing us to conclude that, for $n = 500$ and $n = 1,000$, the HEA algorithm is able to produce the best solutions across the set of all graphs and their isomorphisms under this particular computation limit.

**Table 4.1** Summary of results produced at the computation limit using random graphs $G_{n,0.5}$

| $n$ | TABUCOL | PARTIALCOL | HEA | ANTCOL | HC | Bktr |
|---|---|---|---|---|---|---|
| | | | Algorithm[a] | | | |
| 250 | **28.04± 0.20** | **28.08 ± 0.28** | **28.04 ± 0.33** | 28.56 ± 0.51 | 29.28 ± 0.46 | 34.24 + 0.78 |
| 500 | 49.08 ± 0.28 | 49.24 ± 0.44 | **47.88 ± 0.51** | 49.76 ± 0.44 | 54.52 ± 0.77 | 62.24 ± 0.72 |
| 1000 | 88.92 ± 0.40 | 89.08 ± 0.28 | **85.48 ± 0.46** | 89.44 ± 0.58 | 101.44 ± 0.82 | 112.88 ± 0.97 |

[a] Mean plus/minus standard deviation in number of colours, taken from runs across 25 graphs.

Moving on to other densities, the graphs shown in Figure 4.6 summarise the mean solution quality achieved by the six algorithms on all random graphs generated. In each figure, the bars show the number of colours used in solutions produced by DSATUR and the lines then give the proportion of this number used in the solutions of the six algorithms. Note that all algorithms achieve a reduction in the number of colours realised by DSATUR, though in all but the smallest, sparsest graphs, the backtracking algorithm exhibits the smallest margins of improvement, apparently due to the high levels of vertex homogeneity in these instances, which makes it difficult for favourable regions of the search tree to be identified.

---

[3] As in Chapter 2, statistical significance is claimed here according to the nonparametric Related Samples Wilcoxon Signed Rank test (for pairwise comparisons), and the Related Samples Friedman's Two-way Analysis of Variance by Ranks (for group comparisons). For the remainder of this chapter statistical significance is claimed at the 1% level.

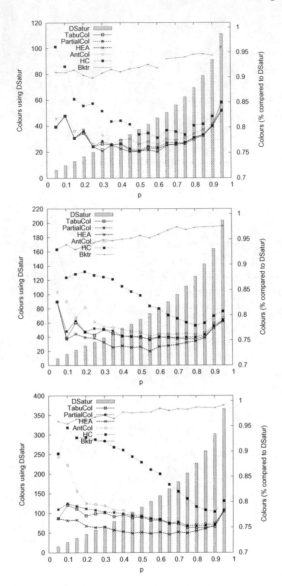

**Fig. 4.6** Mean quality of solution achieved on random graphs using $n = 250$, 500, and 1,000 (respectively) for various edge probabilities $p$. All points are the mean of 25 runs on 25 different instances

It is clear from Figure 4.6 that TABUCOL, PARTIALCOL, and the HEA in particular, produce the best results for the random graphs. For $n = 250$ these algorithms produce mean results that, across the range of values for $p$, show no significant difference among one another, perhaps indicating that the achieved solutions are consistently close to the optimal solutions. For larger graphs however, the HEA's solutions are seen to be significantly better, though its rates of improvement are slightly slower than those of TABUCOL and PARTIALCOL, as illustrated by Figure 4.7. Similar behaviour during runs was also witnessed with the smaller random instances.

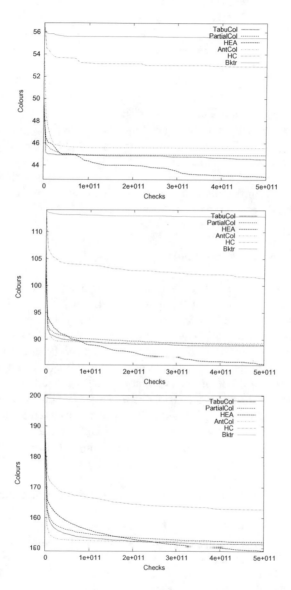

**Fig. 4.7** Run profiles on random graphs of $n = 1,000$ with edge probabilities $p = 0.25, 0.5$, and 0.75 respectively. Each line represents a mean of 25 runs on 25 different instances

Overall, the patterns shown in Figure 4.6 indicate that the HEA's strategy of exploring the space of infeasible solutions using both global and local search operators is the most beneficial of those considered here. Indeed, although the HC algorithm also uses both global and local search operators, here its insistence on preserving feasibility implies a lower level of connectivity in its underlying solution space, making navigation more restricted and resulting in noticeably inferior solutions.

Figure 4.6 also reveals that ANTCOL does not perform well with large sparse instances, though it does become more competitive with denser instances. The reasons for this are twofold. First, the degrees of vertices in sparse graphs are naturally lower, reducing the heuristic bias provided by $\eta$ and perhaps implying an overdominant role of $\tau$ during applications of BUILDSOLUTION (see Equation (4.4)). Secondly, sparse graphs also feature greater numbers of vertices per colour—thus, even if very promising independent sets *are* identified by ANTCOL, their reconstruction by later ants will naturally depend on a longer sequence of random trials, making them less likely. To back these assertions, we also repeated the trials of ANTCOL using the same local search iteration limit as the HEA, $I = 16n$. However, though this brought slight improvements for denser graphs, the results were still observed to be significantly worse than the HEA's, suggesting the difference in performance indeed lies with the global-search element of ANTCOL in these cases.

### 4.2.1.2 Performance on Flat Graphs

Similar patterns are also revealed when we turn our attention towards the performance of the six algorithms with flat graphs, as shown in Figure 4.8. Again, we see that the HEA, TABUCOL, and PARTIALCOL exhibit the best performance on instances within the phase transition regions, with the HC and backtracking algorithms proving the least favourable. One pattern to note is that for the three values of $q$, the HEA tends to produce the best-quality results on the left side of the phase transition region, but PARTIALCOL produces better results for a small range of $p$'s on the right side. However, this difference is not due to the "FOO" tabu tenure mechanism of PARTIALCOL, because no significant difference was observed when we repeated our experiments using PARTIALCOL under TABUCOL's tabu tenure scheme. Thus, it seems that PARTIALCOL's strategy of only allowing solutions to be built from independent sets is favourable in these cases, presumably because this restriction facilitates the formation of independent sets of size $n/q$—structures that will be less abundant in denser graphs, but which also serve as the underlying building blocks in these cases.

Another striking feature of Figure 4.8 is the poor performance of ANTCOL on the right side of the phase transition regions. This again seems to be due to the diminished effect of heuristic value $\eta$, which in this case is due to the variance in vertex degrees being deliberately low, making it difficult to distinguish between vertices. Furthermore, in denser graphs fewer combinations comprising $n/q$ vertices will form independent sets, decreasing the chances of an ant constructing one. This reasoning is also backed by the fact that ANTCOL's poor performance lessens with

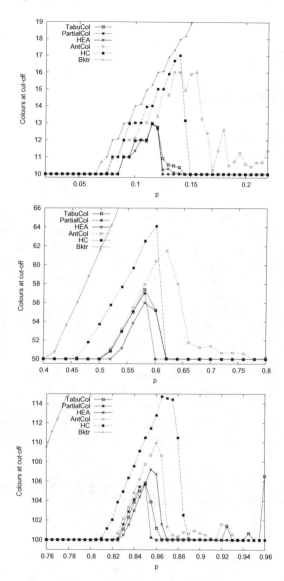

**Fig. 4.8** Mean quality of solution achieved with flat graphs of $n = 500$ with $q = 10$, 50, and 100 (respectively) for various edge probabilities $p$. All points are the mean of 20 runs on 20 different instances

larger values of $q$ where, due to there being fewer vertices per colour, the reproduction of independent sets is dependent on shorter sequences of random trials.

## 4.2.2 Exam Timetabling Problems

Our first set of "real world" problem instances in this comparison concerns graphs representing university timetabling problems. As we saw in Section 1.1.2, timetabling problems involve assigning a set of "events" (exams, lectures, etc.) to a fixed number of "timeslots", and a pair of events "conflict" when they require the same single resource: e.g., there may be a student or lecturer who needs to attend both events, or the events may require use of the same room. As a result conflicting events need to be assigned to different timeslots. Under this constraint, timetabling problems can be modelled as graph colouring problems by considering each event as a vertex, with edges occurring between pairs of events that conflict. Each colour then represents a timeslot, and a feasible colouring corresponds to a complete timetable with no conflict violations.

In practice, universities will often have a predefined number of timeslots in their timetable and their task will be to determine a feasible solution using fewer or equal timeslots. In many cases however, it might be difficult to ascertain whether a timetable with a given number of timeslots is achievable for a particular problem, or it may be desirable to use as few timeslots as possible, particularly if it provides extra time for marking, or allows for a shorter teaching day. Here we concern ourselves with the latter problem, and use a well-known set of real-world timetabling problems compiled by Carter et al. (1996). This set contains 13 exam timetabling problems encountered in various universities from across the globe during the 1980s and 1990s.

| Instance | $n$ | Density | Min;Med;Max | Degree Mean $\mu$ | CV $(\sigma/\mu)$ |
|----------|-----|---------|-------------|-------------------|-------------------|
| hec-s-92 | 81 | 0.415 | 9; 33; 62 | 33.7 | 36.3% |
| sta-f-83 | 139 | 0.143 | 7; 16; 61 | 19.9 | 67.4% |
| yor-f-83 | 181 | 0.287 | 7; 51; 117 | 52 | 35.2% |
| ute-s-92 | 184 | 0.084 | 2; 13; 58 | 15.5 | 69.1% |
| ear-f-83 | 190 | 0.266 | 4; 45; 134 | 50.5 | 56.1% |
| tre-s-92 | 261 | 0.180 | 0; 45; 145 | 47 | 59.6% |
| lse-f-91 | 381 | 0.062 | 0; 16; 134 | 23.8 | 93.2% |
| kfu-s-93 | 461 | 0.055 | 0; 18; 247 | 25.6 | 120.0% |
| rye-s-93 | 486 | 0.075 | 0; 24; 274 | 36.5 | 111.8% |
| car-f-92 | 543 | 0.138 | 0; 64; 381 | 74.8 | 75.3% |
| uta-s-92 | 622 | 0.125 | 1; 65; 303 | 78 | 73.7% |
| car-s-91 | 682 | 0.128 | 0; 77; 472 | 87.4 | 70.9% |
| pur-s-93 | 2419 | 0.029 | 0; 47; 857 | 71.3 | 129.5% |

**Table 4.2** Details of the 13 timetabling instances of Carter et al. (1996)

A summary of these problem instances is provided in Table 4.2. The names of the graphs start with a three-letter code denoting the name of the university. This is followed by an "s" or "f" specifying whether the problem occurred in the summer or fall semester, and this is then followed by the year. We see that the set contains problems ranging in size from $n = 81$ to $2,419$ vertices, and densities of 2.9% up to 41.5%.

It is also known that many of these problem instances feature high numbers of rather large cliques. As Ross et al. (2003) have noted:

> Consider the instance kfu-s-93, by no means the hardest or largest in this set. It involves 5,349 students sitting 461 exams, ideally fitted into 20 timeslots. The problem contains two cliques of size 19 and huge numbers of smaller ones. There are 16 exams that clash with over 100 others.

| Instance | TabuCol | PartialCol | Colours at Cut-off: Mean (best) HEA | AntCol | HC | Bktr |
|---|---|---|---|---|---|---|
| hec-s-92 | 17.22 (17) | 17.00 (17) | 17.00 (17) | 17.04 (17) | 17.00 (17) | 19.00 (19) |
| sta-f-83 | 13.35 (13) | 13.00 (13) | 13.00 (13) | 13.13 (13) | 13.00 (13) | *13.00 (13) [100%, <0.1%] |
| yor-f-83 | 19.74 (19) | 19.00 (19) | 19.06 (19) | 19.87 (19) | 19.00 (19) | 20.00 (20) |
| ute-s-92 | 10.00 (10) | 10.00 (10) | 10.00 (10) | 11.09 (10) | 10.00 (10) | 10.00 (10) |
| ear-f-83 | 26.21 (24) | 22.46 (22) | 22.02 (22) | 22.48 (22) | 22.00 (22) | *22.00 (22) [100%, 0.7%] |
| tre-s-92 | 20.58 (20) | 20.00 (20) | 20.00 (20) | 20.04 (20) | 20.00 (20) | 23.00 (23) |
| lse-f-91 | 19.42 (18) | 17.02 (17) | 17.00 (17) | 17.00 (17) | 17.00 (17) | *17.00 (17) [100%, 1.3%] |
| kfu-s-93 | 20.76 (19) | 19.00 (19) | 19.00 (19) | 19.00 (19) | 19.00 (19) | 19.00 (19) |
| rye-s-93 | 22.40 (21) | 21.06 (21) | 21.04 (21) | 21.55 (21) | **21.00 (21)** | 22.00 (22) |
| car-f-92 | 39.92 (36) | 32.48 (31) | 28.50 (28) | 30.04 (29) | 27.96 (27) | *27.00 (27) [100%, 8.2%] |
| uta-s-92 | 41.65 (39) | 35.66 (34) | 30.80 (30) | 32.89 (32) | 30.27 (30) | **29.00 (29)** |
| car-s-91 | 39.10 (32) | 30.20 (29) | 29.04 (28) | 29.23 (29) | 29.10 (28) | **28.00 (28)** |
| pur-s-93 | 50.70 (47) | 45.48 (42) | 33.70 (33) | 33.47 (33) | 33.87 (33) | **33.00 (33)** |
| **Total** | 341.05 (315) | 302.36 (294) | 280.16 (277) | 286.84 (281) | **279.20 (276)** | 282.00 (282) |
| **Rank** | (6) | (5) | (2) | (4) | (1) | (3) |

**Table 4.3** Summary of algorithm performance on the 13 timetabling instances of Carter et al. (1996). All statistics are collected from 50 runs on each instance. Asterisks in the rightmost column indicate where the backtracking algorithm was able to produce a provably optimal solution. In these cases, the square brackets indicate the % of runs where this occurred, and the average % of the computation limit that this took

Table 4.3 summarises the results achieved at the computation limit with the six graph colouring algorithms. Note that in contrast to the artificial instances from the previous section, the worst overall performance now occurs with those methods relying solely on local search: that is, TabuCol and to a lesser extent PartialCol. Indeed, we find that these methods are often incapable of achieving feasible solutions even using the initial setting for $k$ determined by DSatur.[4] The cause of this poor performance seems to lie in the fact that, as shown in Table 4.2, the degree coefficient of variations (CVs) of these 13 problems are considerably higher than that of the artificially generated instances seen in the previous subsection. The effects

---

[4] Consequently, the reported results for TabuCol and PartialCol in Table 4.3 are produced using an initial $k$ generated by executing the Greedy algorithm with a random permutation of the vertices.

of this are shown in Figure 4.9 where, compared to a random graph of a similar size and density, the differences in cost between neighbouring solutions vary much more widely. This suggests a more "spiky" cost landscape in which the use of local search mechanisms in isolation is insufficient, exhibiting a susceptibility to becoming trapped at local optima.

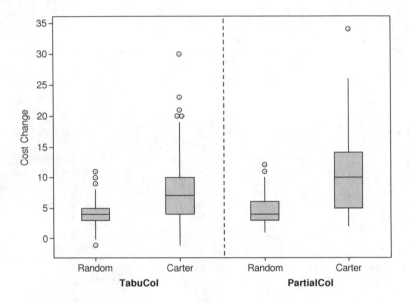

**Fig. 4.9** Cost-change distributions for a random graph ($n = 500$, $p = 0.15$, CV= 10.7%, using $k = 16$) and timetable graph car-f-92 ($n = 543$, $p = 0.138$, CV= 75.3%, using $k = 27$). In all cases samples are taken from candidate solutions with costs of 8

Table 4.3 also shows that the most consistent performance with these graphs is achieved by the HC and HEA algorithms (no significant difference between the two methods across the instances is apparent). This demonstrates that the issues of using local search in isolation are alleviated by the addition of a global search-based operator. On individual instances, the relative performances of HC and HEA do seem to vary, however. With the problem instances car-f-92 and car-s-91, for example, the HEA's best observed solutions are determined within approximately 1% of the computation limit, while HC's progress is much slower. On the other hand, with instances such as rye-s-93, HC consistently produces the best observed results in less that 0.3% of the computation limit, suggesting that its operators are somehow suited to this instance. This issue is considered further in Section 4.4.1.

In contrast to the artificially generated graphs seen earlier, we also observe that the backtracking algorithm is quite competitive with these instances. For four of the problem instances the algorithm has managed to find and prove the optimal solutions in all runs using a small fraction of the computation limit; however, these do not correspond to the smallest instances as we might have expected. In addition, the algorithm has produced the best average performance out of all algorithms with the four *largest* problem instances, seeming to contradict the often-held belief that complete algorithms are only suitable for graphs with less that 100 vertices or so (see Section 3.2). It seems in these cases that the large degree CVs characterise an abundance of heuristic information that is being successfully exploited by the algorithm. Indeed, for the four largest instances, all of the solutions reported in Table 4.3 were actually found in less than 2% of the computation limit, implying that the algorithm quickly arrives at the correct regions of the search tree. However, counterexamples in which the backtracking algorithm consistently produces the worst performance can also be seen in Table 4.3, such as with the smallest instance, hec-s-92.

Finally, we also note the sporadic performance of ANTCOL with these instances. For all but the four largest problems, ANTCOL's best solutions equal those of the other algorithms; however, its averages are less favourable, particularly compared to the HEA and HC algorithms. Consider, for example, the results of ute-s-92 in the table. This problem is consistently solved using ten colours by all methods except ANTCOL, which often requires 11 or 12 colours. In fact, we find that for instances such as these, ANTCOL's performance depends very much on the quality of solutions produced in the first cycle of the algorithm. Due to the low vertex degrees (and reduced influence of $\eta$ that results), Equation (4.4) is predominantly influenced by the pheromone values $\tau$; however, if an 11- or 12-colour solution is produced during the first cycle, features of these suboptimal solutions are still used to update the pheromone matrix $t$, making their reoccurrence in later cycles more likely. The upshot is that ANTCOL is rarely seen to improve upon solutions found in the initial cycle of the algorithm with these instances.

### 4.2.3 Social Networks

Our final set of experiments in this chapter involves graphs representing social networks. Social networks consist of "nodes" (usually individual people) that are "tied" by some sort of inter-dependency such as friendship or belief. Here we consider the social networks of school friends, compiled as part of the USA-based National Longitudinal Study of Adolescent Health project (Moody and White, 2003). The colouring of such networks might be required when we wish to partition the students into groups such that individuals are kept separate from their friends, e.g., for group assignments and team-building exercises (see also Section 1.1.1).

To construct these networks, surveys were conducted in various schools, with each student being asked to list all of his or her friends. In some cases, students were only allowed to nominate friends attending the same school, while in others they could include friends attending a "sister school" (e.g., middle-school students could include friends in the local high school), leading to single-cluster and double-cluster networks respectively. In the resultant graphs, each student is represented by a vertex, with edges signifying a claimed friendship between the associated individuals (see Figure 4.10). Note that in the original data, edges signifying friendships are both directed and weighted; however, in our case directions and weights have been removed to form a simple graph.

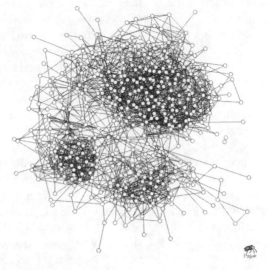

**Fig. 4.10** Illustration of a double-cluster social network collected in the National Longitudinal Study of Adolescent Health project (Moody and White, 2003)

| Instance | n | Density | Min;Med;Max | Degree Mean $\mu$ | CV ($\sigma/\mu$) |
|---|---|---|---|---|---|
| *Single-Cluster* | | | | | |
| #1 | 380 | 0.021 | 0; 8; 23 | 8.1 | 50.5% |
| #2 | 542 | 0.013 | 0; 7; 35 | 7.1 | 61.7% |
| #3 | 563 | 0.013 | 0; 7; 23 | 7.3 | 55.4% |
| #4 | 578 | 0.015 | 0; 8; 24 | 8.8 | 52.7% |
| #5 | 626 | 0.013 | 0; 7; 30 | 7.8 | 58.7% |
| #6 | 746 | 0.010 | 0; 7; 28 | 7.3 | 58.6% |
| #7 | 828 | 0.008 | 0; 6; 23 | 6.2 | 59.3% |
| #8 | 877 | 0.009 | 0; 7; 29 | 7.8 | 58.2% |
| #9 | 1229 | 0.003 | 0; 4; 17 | 4.1 | 54.6% |
| #10 | 2250 | 0.002 | 0; 4; 25 | 4.3 | 78.0% |
| *Double-Cluster* | | | | | |
| #11 | 291 | 0.027 | 0; 8; 21 | 7.8 | 54.6% |
| #12 | 426 | 0.018 | 0; 7; 26 | 7.5 | 56.2% |
| #13 | 457 | 0.016 | 0; 7; 23 | 7.4 | 58.8% |
| #14 | 495 | 0.017 | 0; 8; 22 | 8.5 | 46.8% |
| #15 | 569 | 0.017 | 0; 9; 34 | 9.4 | 50.9% |
| #16 | 586 | 0.016 | 0; 9; 30 | 9.6 | 48.4% |
| #17 | 689 | 0.010 | 0; 6; 22 | 6.8 | 62.0% |
| #18 | 795 | 0.011 | 0; 9; 24 | 8.7 | 53.7% |
| #19 | 1089 | 0.007 | 0; 8; 29 | 8.1 | 57.9% |
| #20 | 1246 | 0.007 | 0; 9; 33 | 8.6 | 54.4% |

**Table 4.4** Details of the 20 social networks used

In our tests we took a random sample of ten single-cluster networks and ten double-cluster networks from the Adolescent Health data set. Summary statistics of these graphs are given in Table 4.4. These figures indicate that the vertex degrees are far lower that the timetabling graphs from the previous section, with the highest degree across the whole set being just 29. Consequently, the densities of the graphs are also much lower.

| Instance | Colours at Cut-off: Mean (best) | | | | | |
|---|---|---|---|---|---|---|
| | TABUCOL | PARTIALCOL | HEA | ANTCOL | HC | Bktr |
| *Single-Cluster* | | | | | | |
| #1 | 8 (8) | 8 (8) | 8 (8) | 8.15 (8) | 8 (8) | 8 (8) |
| #2 | 6 (6) | 6 (6) | 6 (6) | 6.76 (6) | 6 (6) | *6 (6) [100%, <1%] |
| #3 | 7 (7) | 7 (7) | 7 (7) | 7.45 (7) | 7 (7) | 7.02 (7) |
| #4 | 8 (8) | 8 (8) | 8 (8) | 8.75 (8) | 8 (8) | 8 (8) |
| #5 | 8 (8) | 8 (8) | 8 (8) | 8.41 (8) | 8 (8) | 8 (8) |
| #6 | 6 (6) | 6 (6) | 6 (6) | 6 (6) | 6 (6) | *6 (6) [90%, <1%] |
| #7 | 6 (6) | 6 (6) | 6 (6) | 6.38 (6) | 6 (6) | 6 (6) |
| #8 | 8 (8) | 8 (8) | 8 (8) | 8.23 (8) | 8 (8) | 8 (8) |
| #9 | 6 (6) | 6 (6) | 6 (6) | 6.10 (6) | 6 (6) | 6 (6 |
| #10 | 5 (5) | 5 (5) | 5 (5) | 5 (5) | 5 (5) | *5.38 (5) [52%, <1%] |
| *Double-Cluster* | | | | | | |
| #11 | 6 (6) | 6 (6) | 6 (6) | 6.70 (6) | 6 (6) | 6.02 (6) |
| #12 | 5 (5) | 5 (5) | 5 (5) | 5 (5) | 5 (5) | *5 (5) [96%, 4%] |
| #13 | 6 (6) | 6 (6) | 6 (6) | 6 (6) | 6 (6) | *6.32 (6) [46%, 1%] |
| #14 | 7 (7) | 7 (7) | 7 (7) | 7.46 (7) | 7 (7) | *7 (7) [42%, <1%] |
| #15 | 7 (7) | 7 (7) | 7 (7) | 7 (7) | 7 (7) | *7 (7) [100%, <1%] |
| #16 | 10 (10) | 10 (10) | 10 (10) | 10.13 (10) | 10 (10) | 10 (10) |
| #17 | 7 (7) | 7 (7) | 7 (7) | 7.28 (7) | 7 (7) | 7 (7) |
| #18 | 6 (6) | 6 (6) | 6 (6) | 6 (6) | 6 (6) | *6.14 (6) [86%, 1%] |
| #19 | 7 (7) | 7 (7) | 7 (7) | 7.65 (7) | 7 (7) | 7.13 (7) |
| #20 | 7 (7) | 7 (7) | 7 (7) | 7.69 (7) | 7 (7) | 7.02 (7) |
| **Total** | 136 (136) | 136 (136) | 136 (136) | 142.14 (136) | 136 (136) | 137.03 (136) |
| **Rank** | (1) | (1) | (1) | (6) | (1) | (5) |

**Table 4.5** Summary of algorithm performance on the 20 Social Networks. All statistics are collected from 50 runs on each instance. Asterisks in the rightmost column indicate where the backtracking algorithm was able to produce a provably optimal solution. In these cases, the square brackets indicate the % of runs where this occurred, and the average % of the computation limit that this took

As before, each algorithm was executed 50 times on each instance. The relatively straightforward outcomes of these trials are summarised in Table 4.5. Here, we see that the number of colours needed for these problems ranges from five to ten, though no obvious correlations exist to suggest any links with instance size, density, or the presence of clusters. We also see that the HEA, HC, TABUCOL, and PARTIALCOL methods have all produced the best observed (or optimal) solutions for all instances in all runs. It seems, therefore, that the underlying structures and relative sparsity of these graphs make their solving relatively "easy" with these algorithms.

In addition, for six of the instances, the backtracking algorithm has managed to find provably optimal solutions, though this does not occur in all runs. Indeed, when this does happen, it seems to occur early in the process (<5% of the computation limit), suggesting that the random elements of the algorithm can have a large effect on the structure of the search tree. We also observe the poor performance of

ANTCOL, which seems to be due to the negative performance features noted in the previous subsection, with a high-quality solution either being produced very quickly (in the first cycle), or not at all.

## 4.3 Conclusions

As we have seen, the results of the comparison above reveal a complicated picture, with different algorithms outperforming others on different occasions. This suggests that the underlying structures of graphs are often critical in an algorithm's resultant performance. In terms of overall patterns we offer the following observations:

- Algorithms that rely solely on local search (in this case TABUCOL and PAR-TIALCOL) often struggle with instances whose cost landscapes are "spiky", commonly characterised by high coefficient of variations (CVs) in vertex degrees. On the other hand, these methods do show more promise when the degree CV is quite low, such as with random and flat graphs, suggesting that they have a natural aptitude for navigating spaces in which neighbouring solutions feature costs that are often close or equal to that of the incumbent.
- One obvious advantage of the backtracking algorithm is its ability to produce provably optimal solutions. This has occurred for a number of the real-world problem instances considered in our trials, including some rather large instances; however, predicting when this will happen seems difficult. For graphs that are more "regular" in structure, such as the random and flat instances, the performance of the backtracking algorithm is also significantly worse than that of the other approaches.
- Across the trials, HEA has proved to be by far the most consistent of the six approaches. We suggest this is due to a combination of the following attributes:
  - *The HEA operates in the space of infeasible solutions.* Unlike the HC algorithm, which only permits changes to a solution that maintain feasibility, the strategy of allowing infeasible solutions seems to offer higher levels of connectivity (and thus less restriction of movement) within the solution space, helping the algorithm to navigate its way towards high-quality solutions more effectively.
  - *The HEA makes use of global as well as local search operators.* On many occasions TABUCOL performs poorly when used in isolation; however, the HEA's use of global search operators in conjunction with TABUCOL seems to alleviate these problems by allowing the algorithm to regularly escape from local optima.
  - *The HEA's global search operators are robust.* Unlike ANTCOL's global search operator, which sometimes hinders performance, the HEA's use of recombination in conjunction with a small population of candidate solutions seems beneficial across the instances. This is despite the fact that across all of our tests, recombination was never seen to consume more than 2% of the

available run time. Note in particular that the GPX operator does not consider any problem-specific information in its operations (such as the connectivity or degree of vertices), yet it still seems to strike a useful balance between (a) altering the solution sufficiently, while (b) maintaining useful substructures formed in earlier iterations of the algorithm.

One of our intentions in this comparison has been to test the robustness of our six algorithms by executing them blindly on each problem instance. As we have seen, this has involved using the same parameter values (or methods for calculating them) across all trials. However, it should be noted that different settings may lead to better results in some cases. A broader issue concerns how me might go about reliably predicting the performance of a particular graph colouring algorithm on a previously unseen graph. Accurate predictions would obviously be useful here because, given a particular graph, we would then be able to apply the most appropriate method from the available portfolio of algorithms. Work in this area with this chapter's six algorithms has been carried out by Smith-Miles et al. (2014), who use machine learning to classify the types of graph that the different algorithms are seen to perform well with. This information is then used to help predict the best performing algorithm on future unseen problem instances.

Finally, as with Chapter 2, this chapter's comparison has been carried out using a platform independent measure of computational effort. In terms of CPU time, Table 4.6 shows the relative run times of the algorithms for a small sample of graphs. Perhaps the most striking feature is that the HEA is among one of the quickest to execute, a fact that further endorses the method. On the other hand, the ANTCOL and the HC algorithms seem to require significantly more time, apparently due to the computational overheads associated with their BUILDSOLUTION and Kempe chain operators respectively.

|             | $n = 250$ | 500  | 1000 |
|-------------|-----------|------|------|
| TABUCOL     | 1346      | 1622 | 1250 |
| PARTIALCOL  | 1435      | 1372 | 1356 |
| HEA         | 1469      | 1400 | 1337 |
| ANTCOL      | 4152      | 3840 | 4349 |
| HC          | 5829      | 5473 | 5320 |
| Bktr        | 6328      | 4794 | 3930 |

**Table 4.6** Time (in seconds) to complete runs of $5 \times 10^{11}$ constraint checks with random graphs $G_{n,0.5}$ using a 3.0 GHz Windows 7 PC with 3.87 GB RAM

## 4.4 Further Improvements to the HEA

We now conclude this chapter by looking at some of the individual elements of the HEA and proposing some ideas as to how its performance might be further improved in some cases.

### 4.4.1 Maintaining Diversity

In general, an important factor behind the behaviour of an evolutionary algorithm (EA) is the level of diversity that is maintained within its population during a run. Typically, in early iterations of an EA the diversity of a population will be high, allowing the algorithm to consider many different parts of the solution space. This is often known as the "exploration" phase of the algorithm. As the population evolves over time this diversity then generally falls as the algorithm "homes in" on promising regions of the solution space and seeks to search within these areas more thoroughly. This is often called the "exploitation" phase.

It is clear that when applying an EA to a computational problem, a suitable balance will need to be established between exploration and exploitation. A fall in diversity that is too slow is undesirable because the algorithm will devote too much energy into broadly scanning the whole solution space, as opposed to intensively searching specific regions within it. On the other hand, a fall in diversity that is too rapid can also be problematic because the EA will spend too much time focussing on limited regions of the solution space. The latter issue is often called *premature convergence*.

To examine the issue of diversity with the HEA for graph colouring, let us first define a metric for measuring the distance between two candidate solutions.

**Definition 4.1** *Given a solution $\mathcal{S}$, let $\mathcal{P}_\mathcal{S} = \{\{u,v\} : u,v \in V \wedge u \neq v \wedge c(u) = c(v)\}$. The distance between two solutions $\mathcal{S}_1$ and $\mathcal{S}_2$ can then be evaluated using the Jaccard distance measure on the sets $\mathcal{S}_1$ and $\mathcal{S}_2$. That is:*

$$D(\mathcal{S}_1, \mathcal{S}_2) = \frac{|\mathcal{P}_{\mathcal{S}_1} \cup \mathcal{P}_{\mathcal{S}_2}| - |\mathcal{P}_{\mathcal{S}_1} \cap \mathcal{P}_{\mathcal{S}_2}|}{|\mathcal{P}_{\mathcal{S}_1} \cup \mathcal{P}_{\mathcal{S}_2}|}. \tag{4.8}$$

This distance measure gives the proportion of vertex pairs (assigned to the same colour) that exist in just one of the two solutions. Consequently, if the solutions $\mathcal{S}_1$ and $\mathcal{S}_2$ are identical, then $\mathcal{P}_{\mathcal{S}_1} \cup \mathcal{P}_{\mathcal{S}_2} = \mathcal{P}_{\mathcal{S}_1} \cap \mathcal{P}_{\mathcal{S}_2}$, giving $D(\mathcal{S}_1, \mathcal{S}_2) = 0$. Conversely, if no vertex pair is assigned the same colour, $\mathcal{P}_{\mathcal{S}_1} \cap \mathcal{P}_{\mathcal{S}_2} = \emptyset$, implying $D(\mathcal{S}_1, \mathcal{S}_2) = 1$.

Given this distance measure, we are also able to define a population diversity metric. This is defined as the mean distance between each pair of solutions in the population.

**Definition 4.2** *Given a population of solutions defined as a multiset $\mathbf{S} = \{\mathcal{S}_1, \mathcal{S}_2, \ldots, \mathcal{S}_{|\mathbf{S}|}\}$, the diversity of $\mathbf{S}$ is calculated:*

$$\text{Diversity}(\mathbf{S}) = \frac{1}{\binom{|\mathbf{S}|}{2}} \sum_{\forall \mathcal{S}_i, \mathcal{S}_j \in \mathbf{S}: i < j} D(\mathcal{S}_i, \mathcal{S}_j). \tag{4.9}$$

When applying the HEA to the graphs considered in this chapter, we found that satisfactory levels of diversity were maintained in most cases. However, for some of the timetabling problem instances we also observed that large colour classes of low-degree vertices are often formed in early stages of the algorithm, and that these quickly come to dominate the population, causing premature convergence. Indeed, as can be seen in Table 4.3, the HEA often produces inferior results with these problems.

One method by which population diversity might be prolonged in EAs is to make larger changes (mutations) to an offspring in order to increase its distance from its parents. However, this must be used with care, particularly because changes that are too large might significantly worsen a solution, undoing much of the work carried out in previous iterations of the algorithm. For the HEA, one obvious way of making more changes to a solution is to increase the iteration limit of the local search procedure. However, although this might allow further improvements to be made to a solution, it might also slow the algorithm unnecessarily.

An alternative method for maintaining diversity might be to alter the HEA's recombination operator so that it works exclusively with proper colourings. As noted in Section 4.1.3, the GPX operator considers candidate solutions in which clashes are permitted; however, in practice this could allow large colour classes containing clashes to be unduly promoted in the population, when perhaps the real emphasis should be on the promotion of large *independent sets*. Consequently, we might refine the GPX operator by first removing all clashing vertices from each parent before performing recombination. This implies that, before assigning missing vertices to random colours, an offspring will always be proper. A further effect is that a greater number of vertices will usually need to be recoloured because the vertices originally removed from the parents may also be missing in the resultant offspring. Hence the resultant offspring will tend to be less similar to its parents.

If the above option is chosen, then before randomly reassigning missing vertices to colours, we also have the opportunity to alter the partial proper solution using Kempe chain interchanges. Recall from Theorem 3.1 that this operator, when applied to a proper solution, does not introduce any clashes. Thus we are provided with a mechanism by which we can make changes to a solution without compromising its quality in any way.

To illustrate the potential effects of this latter scheme, Figure 4.11 shows the levels of diversity that exist in the HEA's population for the first 3,000 iterations of a run using the timetabling graph car-s-91, which has a chromatic number of 28. When using the original HEA, the population has converged at around 500 iterations and, as we saw in Table 4.3, the algorithm produces solutions using more than 29 colours on average. On the other hand, by applying a series of random Kempe chain moves ($2k$ moves per iteration in this case), population diversity is maintained at a much higher level. In our tests this modification enabled the algorithm to determine optimal 28-colourings in all runs.

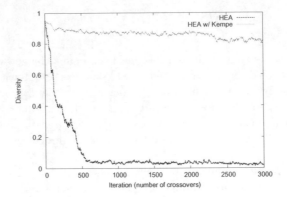

**Fig. 4.11** Population diversity using a population of size 10 with the timetabling problem instance car-s-91, using $k = 28$

We should note that using Kempe chain interchanges in this way is not always beneficial, however. For instance, similar tests to the above were also carried using random and flat graphs. When using a suitably low value for $k$ in these cases, we found that the Kempe chain interchange operator was usually unable to alter the underlying structures of solutions because its application nearly always resulted in colour relabellings (or in other words, the bipartite graphs induced by each pair of colour classes in these solutions were nearly always connected, giving total Kempe chains). As an aside, it would be interesting to investigate whether this property in a solution gives any clues as to how close it is to the optimal solution.

Note that within this book's suite of graph colouring algorithms, the HEA contains run-time options for outputting the population diversity and for applying Kempe chain interchanges in the manner described above. (Refer to the algorithm user guide in Appendix A.1 for further information.)

### 4.4.2 Recombination

Since the proposal of the GPX by Galinier and Hao (1999), further recombination operators based on their scheme have also been suggested, differing primarily on the criteria used for deciding which colour classes to copy to the offspring. Porumbel et al. (2010), for example, suggest that instead of choosing the largest available colour class at each stage of the recombination process, classes with the *least number of clashes* should be prioritised, with class size and information regarding the degrees of the vertices then being used to break ties. Lü and Hao (2010a), on the other hand, have proposed extending the GPX operator to allow more than two parents to play a part in producing a single offspring. In this multi-parent operator, offspring are constructed in the same manner as the GPX, except that at each stage the largest colour class from *multiple parents* is chosen to be copied into the off-spring. The intention behind this increased choice is that larger colour classes will be identified, resulting in fewer uncoloured vertices once the $k$ colour classes have been constructed. In order to prohibit too many colours being inherited from one particular parent, the authors make use of a parameter $q$, specifying that if the $i$th colour class in an offspring is copied from a particular parent, then this parent should not be considered for a further $q$ colours. Note that GPX is simply an application of this operator using two parents with $q = 1$.

Another method of recombination with the graph colouring problem involves considering the individual assignments of vertices to colours as opposed to their partitions. Here, a natural way of representing a solution is to use a vector $(c(v_1), c(v_2), \ldots, c(v_n))$, where $c(v_i)$ gives the colour of vertex $v_i$. However, it has long been argued that this sort of approach has disadvantages, not least because it leads to a solution space that is far larger than it needs to be, since any solution using $k$ colours can be represented in $k!$ ways (refer to Section 1.4.1). Furthermore, authors such as Falkenauer (1998) and Coll et al. (1995) have also argued that "traditional" recombination schemes such as 1-, 2-, and $n$-point crossover with this representation have a tendency to recklessly break up building blocks that we might want promoted in a population.

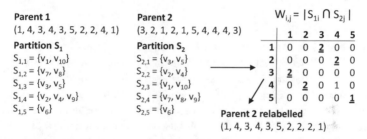

**Fig. 4.12** Example of the relabelling procedure proposed by Coll et al. (1995). Here, parent 2 is relabelled $1 \to 3$, $2 \to 4$, $3 \to 1$, $4 \to 2$, and $5 \to 5$

In recognition of the perceived disadvantages of the assignment-based representation, Coll et al. (1995) have proposed a procedure for relabelling the colours of one of the parents before applying one of these "traditional" crossover operators. Consider two (not necessarily feasible) parent solutions represented as partitions: $S_1 = \{S_{1,1}, \ldots, S_{1,k}\}$ and $S_2 = \{S_{2,1}, \ldots, S_{2,k}\}$. Now, using $S_1$ and $S_2$, a complete bipartite graph $K_{k,k}$ is formed. This bipartite graph has $k$ vertices in each partition, and the weights between two vertices from different partitions is defined as $W_{i,j} = |S_{1,i} \cap S_{2,j}|$. Given $K_{k,k}$, a maximum weighted matching can then be determined using any suitable algorithm (such as the Hungarian algorithm (Munkres, 1957) or Auction algorithm (Bertsekas, 1992)), and this matching can be used to relabel the colours in one of the parents. Figure 4.12 gives an example of this procedure and shows how the second parent can be altered so that its colour labellings maximally match those of the first parent. In this example we see that the colour classes $\{v_1, v_{10}\}$, $\{v_3, v_5\}$, and $\{v_6\}$ occur in both parents and will be preserved in any offspring produced via a traditional operator such as uniform crossover. However, this will not always be the case and will depend very much on the best matching available in each case.

An interesting point regarding the structure of solutions and the resultant effects of recombination has also been raised by Porumbel et al. (2010). Specifically, they propose that when solutions involve a small number of large colour classes (such as solutions to sparse graphs), good quality solutions will tend to occur through the identification of large independent sets, perhaps suggesting that the GPX and its multi-parent variant are naturally suited in these cases. On the other hand, if a solution involves many small colour classes, quality seems to be determined more through the identification of good *combinations* of independent sets.

| Parent $S_1$ | Parent $S_2$ | Offspring $S'$ |
|---|---|---|
| $S_{1,1} = \{v_1, v_{10}\}$ | $S_{2,1} = \{v_1, v_9\}$ | $S'_1 = \{v_1, v_{10}\}$ |
| $S_{1,2} = \{v_7, v_8\}$ | $S_{2,2} = \{v_7\}$ | $S'_2 = \{v_7, \not{v_8}\}$ |
| $S_{1,3} = \{v_3, v_5\}$ | $\blacktriangleright S_{2,3} = \{v_3, v_5\}$ | $S'_3 = \{v_3, v_5\}$ |
| $S_{1,4} = \{v_2, v_4, v_9\}$ | $\blacktriangleright S_{2,4} = \{v_2, v_4, v_8\}$ | $S'_4 = \{v_2, v_4, v_8\}$ |
| $S_{1,5} = \{v_6\}$ | $S_{2,5} = \{v_6, v_{10}\}$ | $S'_5 = \{v_6\}$  Uncoloured $= \{v_9\}$ |

**Fig. 4.13** Demonstration of the GGA recombination operator. Here, the colour classes in parent 2 have been labelled to maximally match those of parent 1

To these ends a further recombination operator for graph colouring is also proposed by Lewis (2015) which, unlike GPX, shows no bias towards offspring inheriting larger colour classes, or towards offspring inheriting half of its colour classes from each parent. An example of this operator is given in Figure 4.13. Given two parents, the colour classes in the second parent are first relabelled using Coll et al.'s procedure from above. Using the partition-based representations of these solutions, a subset of colour classes from the second parent is then selected randomly, and these replace the corresponding colours in a copy of the first parent. Duplicate vertices are then removed from colour classes originating from the first parent, and any uncoloured vertices are assigned to random colour classes. Tests by Lewis (2015) indicate that this recombination operator can produce marginally better solutions than the GPX operator when colour classes are small (approximately five vertices per colour), though worse results are also seen to occur in other cases.

Note that the recombination operators listed in this subsection are also included as run-time options within this book's suite of graph colouring algorithms (see Appendix A.1).

### 4.4.3 Local Search

Finally, from the analyses in this chapter it is apparent that graph colouring algorithms such as the HEA usually benefit when used in conjunction with an appropriate local search procedure. For algorithms operating in the space of complete improper solutions, this has often been provided by the TABUCOL algorithm. The tabu search metaheuristic seems to be very suitable for this purpose because, by extending the steepest descent algorithm, it allows rapid improvements to be made to a solution.

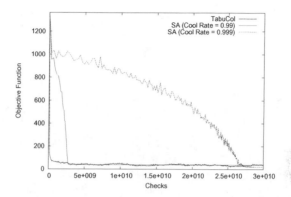

**Fig. 4.14** Example run profiles of TABUCOL and an analogous SA algorithm using a random graph $G_{1000,0.5}$ with $k = 86$. Here the SA algorithm uses an initial value for $t = 0.7$, with $z = 500,000$

To contrast this, consider the rates of improvement achieved by an analogous simulated annealing algorithm that uses the same neighbourhood operator as TABU-COL but which follows the pseudocode given in Figure 3.5. For this algorithm, values need to be determined for the initial temperature $t$, the cooling rate $\alpha$, and the frequency of temperature updates $z$. Figure 4.14 compares the run profile of TABUCOL to this simulated annealing algorithm on an example random graph. It can be seen that TABUCOL quickly reduces the objective function (Equation (4.1)), while the SA approach takes much longer. In addition, the SA algorithm seems quite sensitive to adjustments in its parameters, with inappropriate values potentially hindering performance. On the other hand, it is well known that when the temperature is reduced more slowly, runs of SA tend to produce better quality solutions (van Laarhoven and Aarts, 1987). Hence, with extended run times SA may have the potential to produce superior solutions to TABUCOL in some cases.

# Chapter 5
# Applications and Extensions

We are now at a point in this book where we have seen a number of different algorithms for the graph colouring problem and have noted many of their relative strengths and weaknesses. In this chapter we now present a range of problems, both theoretical- and practical-based, for which such algorithms might be applied including face colouring, edge colouring, pre-colouring, solving Latin squares and sudoku puzzles and testing for short circuits on printed circuit boards. Note that these problems actually represent special cases of the general graph colouring problem in that their underlying graphs conform to specific topologies, as we shall see.

This chapter also considers a variant of the graph colouring problem where not all of the graph is visible to an algorithm, or where the graph is subject to change over time. Such problems can arise when setting up wireless ad hoc networks and also in some timetabling applications. We then go on to consider problems that *extend* and therefore generalise the graph colouring problem, specifically list colouring, equitable colouring and weighted graph colouring. Detailed real-world applications of graph colouring are also the subject of Chapters 6, 7 and 8.

Note that, in contrast to the rest of this book, the first two sections of this chapter are concerned with colouring the *faces* of graphs and the *edges* of graphs as opposed to the vertices. As we will see, the latter two problems can be converted into equivalent formulations of the vertex colouring problem using the concepts of *dual graphs* and *line graphs* respectively. However, it is often useful for face and edge colouring problems to be considered as separate problems; hence we will often use the term "vertex colouring" instead of "graph colouring" to avoid any ambiguities.

## 5.1 Face Colouring

In the face colouring problem we want to colour the *spaces* between vertices and edges, as opposed to the vertices themselves. Face colouring is specifically concerned with planar graphs which, as we saw in Chapter 1, are graphs that can be

© Springer International Publishing Switzerland 2016
R.M.R. Lewis, *A Guide to Graph Colouring*,
DOI 10.1007/978-3-319-25730-3_5

drawn on a plane so that no edges cross one another. When drawn in this way planar graphs can be divided into faces, including one unbounded face that surrounds the graph. Figure 5.1, for example, shows a planar graph comprising ten faces: nine bounded faces and one unbounded face (numbered 10 in the figure). The *boundary* of a face is the set of edges that surrounds it and, when the face is bounded, these edges form a cycle.

It is evident by inspecting Figure 5.1 that the number of faces seems to be related to the number of vertices and edges of the graph. In fact, this relationship can be stated explicitly due to an elegant theorem that was first noted in the 1700s by mathematician Leonhard Euler (1707–1783):

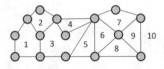

**Fig. 5.1** Planar graph with $n = 15$ vertices, $m = 23$ edges and $f = 10$ faces

**Theorem 5.1 (Euler's characteristic)** *Let G be a planar graph with n vertices, m edges, and f faces. Then*

$$n - m + f = 2$$

*Proof.* The proof is via induction on the number of faces $f$. If $f = 1$, then the graph contains no cycles and must therefore be a tree. Since the number of edges in a tree $m = n - 1$, the theorem holds because $n - (n - 1) + 1 = 2$.

Now assume $f \geq 2$, meaning that $G$ must now contain at least one cycle. Let $\{u, v\}$ be an edge in one of these cycles. Since this cycle divides two faces, say $F_1$ and $F_2$, removing $\{u, v\}$ from $G$ to form a subgraph $G'$ will have the effect of joining $F_1$ and $F_2$ with all other faces remaining unchanged. Hence $G'$ has $f - 1$ faces.

Let $n'$, $m'$, and $f'$ be the number of vertices, edges, and faces in $G'$. Thus, $n' = n$, $m' = m - 1$, $f' = f - 1$, and $n - m + f = n' - m' + f' = 2$.                                        □

We see that Euler's characteristic does indeed hold for the example graph in Figure 5.1 since $n - m + f = 15 - 23 + 10 = 2$ as required.

When considering the face colouring problem it is necessary to restrict ourselves to planar graphs that contain no *bridges*. A bridge is defined as an edge in a graph $G$ whose removal increases the number of components. When a graph contains a bridge $\{u, v\}$, the unbounded face will surround the graph, but will also feature $\{u, v\}$ on its boundary twice, making it impossible to colour. Hence planar graphs containing bridges are not considered further in this section.

Let us now consider the maximum number of edges that a graph can feature while retaining the property of planarity. Consider a connected planar graph $G$ with $n$ vertices, $m$ edges, and $f$ faces. Also write $f_i$ for the number of faces in $G$ that are surrounded by exactly $i$ edges in their boundaries. Clearly $\sum_i f_i = f$ and, assuming that $G$ does not contain a bridge,

$$\sum_i = if_i = 2m \tag{5.1}$$

since every edge is on the boundary of exactly two faces. We can use this relationship in conjunction with Euler's characteristic to give an upper bound on the number of edges in a planar graph. This result also involves knowledge of the *girth* of a graph, defined as follows.

**Definition 5.1** *The* girth *of a graph G is the length of the shortest cycle in G. If G is acyclic (i.e., contains no cycles), then its girth equals infinity.*

**Theorem 5.2** *Let G be a planar graph with $n \geq 3$ vertices, m edges, f faces and no bridges. Then G has at most $3n - 6$ edges. Furthermore, if G has a girth g (for $3 \leq g \leq \infty$), then:*

$$m \leq \max \left\{ \frac{g}{g-2}(n-2), n-1 \right\}$$

*Proof.* For $g = 3$, we get $\max\{\frac{3}{3-2}(n-2), n-1\} = 3(n-2)$, giving $m \leq 3n - 6$ as required. Hence we only need to prove the second assertion above.

If $g > n$, then this implies $g = \infty$ meaning that G has no cycles and is therefore a tree. Hence, $m = (n-1) \leq n$. Now assume that $g \leq n$ and that the assertion holds for smaller $n$'s. Also assume without loss of generality that G is connected. From earlier we know that:

$$2m = \sum_i if_i = \sum_{i \geq g} if_i \geq g \sum_i f_i = gf.$$

Hence by Euler's characteristic (Theorem 5.1), we get:

$$m \mid 2 = n + f \leq n + \frac{2}{g}m$$

and so

$$m \leq \frac{g}{g-2}(n-2)$$

as required. □

Theorem 5.2 can often be used to decide whether a graph is planar or not. For example, the complete graph with five vertices $K_5$ cannot possibly be planar because it has $n = 5$ vertices and $m = 10$ edges, meaning $m \leq 3n - 6$ is not satisfied. As another example, the complete bipartite graph with six vertices $G = (V_1, V_2, E)$, where $E = \{\{v, u\} : v \in V_1, u \in V_2\}$ and $|V_1| = |V_2| = 3$, is also not bipartite since it has $m = 9$ edges, $n = 6$ vertices, and a girth of 4, meaning that $m - 9 > \frac{4}{4-2}(6-2) = 8$. Less obvious, but profoundly more useful, however, is the amazing fact that a graph is planar *if and only if* it does not contain a subgraph that is a subdivision of either of these two examples. This result, due to Kuratowski (1930), has been used alongside similar results to help construct a number of efficient (polynomial-time) algorithms for determining whether a graph is planar or not, including the Path Addition method of Hopcroft and Tarjan (1974) and the more recent Edge Addition method of Boyer (2004).

### 5.1.1 Dual Graphs, Colouring Maps, and the Four Colour Theorem

The close relationship between the problems of vertex colouring and face colouring becomes apparent when we consider the concept of *dual graphs*. Given a planar graph $G$, the *dual* of $G$, denoted by $G^*$, is constructed according to the following steps. First, draw a single vertex $v_i^*$ inside each face $F_i$ of $G$. Second, for each edge $e$ in $G$, draw a line $e^*$ that crosses $e$ but no other edge in $G$, and that links the two vertices in $G^*$ corresponding to the two faces in $G$ that $e$ is separating.

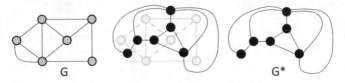

**Fig. 5.2** Illustration of how to convert a planar graph $G$ to its dual $G^*$

This procedure is demonstrated in Figure 5.2. Here, the vertices in $G$ are shown in grey, and the vertices in $G^*$ are shown in black. $G$ has six faces in total, five bounded faces and one unbounded face. The unbounded face is represented by the top vertex of $G^*$ in the example and is made adjacent to all vertices in $G^*$ whose corresponding faces in $G$ have an edge on the exterior of the graph. Note that $G^*$ may also have multiple edges between a pair of vertices, as is occurring on the right-hand side of the example graph.

It is clear from the figure that the process of forming duals is reversible: that is, we can use the same process to form $G$ from $G^*$. It is also clear that because $G$ is planar, its dual $G^*$ must also be planar. We are also able to state simple relationships between the number of vertices, faces and edges in $G$ and $G^*$ such as the following:

**Theorem 5.3** *Let $G$ be a connected planar graph with $n$ vertices, $m$ edges, and $f$ faces. Also, let $G^*$ be the dual of $G$, comprising $n^*$ vertices, $m^*$ edges, and $f^*$ faces. Then $n^* = f$, $m^* = m$ and $f^* = n$.*

*Proof.* It is clear that $n^* = f$ due to the method by which duals are constructed. Similarly $m^* = m$ because all edges in $G^*$ intersect exactly one edge each in $G$ (and vice versa). The third relation follows by substituting the previous two relationships into Euler's characteristic applied to both $G$ and $G^*$.                                     □

Recall from Chapter 1 that the Four Colour Theorem (or "conjecture" as it was at the time) was originally stated in 1852 by Francis Guthrie, who hypothesised that four colours are sufficient for colouring the faces of any map such that neighbouring faces have different colours. In the context of graph theory, a map can be

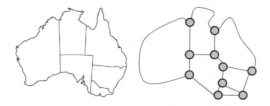

**Fig. 5.3** The territories of mainland Australia (left), and the corresponding planar graph (right)

represented by a bridge-free planar graph $G$, with the faces of $G$ representing the various regions of the map, edges representing borders between regions, and vertices representing points where the borders intersect. An illustration using a map of Australia (excluding the Australian Capital Territory) is given in Figure 5.3.

The following theorem now reveals the close relationship between the vertex colouring and face colouring problems:

**Theorem 5.4** *Let $G$ be a connected planar graph without loops, and let $G^*$ be its dual. Then the vertices of $G$ are $k$-colourable if and only if the faces of $G^*$ are $k$-colourable.*

*Proof.* Since $G$ is connected, planar, and without loops, its dual $G^*$ is a planar graph with no bridges. If we have a $k$-colouring of the vertices of $G$, then each face of $G^*$ can now be assigned to the same colour as its corresponding vertex in $G$. Because no adjacent vertices in $G$ have the same colour, it follows that no adjacent faces in $G^*$ will have the same colour. Thus the faces of $G^*$ are $k$-colourable.

Now suppose that we have a $k$-colouring of the faces of $G^*$. Since every vertex of $G$ is contained in a face of $G^*$, each vertex in $G$ can assume the colour of its corresponding face in $G^*$. Again, since no adjacent faces in $G^*$ are allocated the same colour, this implies no adjacent vertices in $G$ are given the same colour. □

This result is important because it tells us that the faces of any map, represented as a planar graph $G^*$ with no bridges, can be $k$-coloured by simply determining a vertex $k$-colouring of its dual graph $G$. The result also tells us that we can take any theorem concerning the vertex colouring of a planar graph and then state a corresponding theorem on the face colouring of its dual, and vice versa. One elegant theorem that arises from this characteristic demonstrates a link between a special type of topology known as Eulerian graphs and graphs that are bipartite.

**Definition 5.2** *A graph $G$ is* Eulerian *if and only if the degree of all vertices in $G$ are even.*

This gives rise to the following theorem:

**Theorem 5.5** *The faces of a planar graph with no bridges G are 2-colourable if and only if G is Eulerian.*

*Proof.* Recall that a graph's vertices are 2-colourable if and only if it is bipartite. Hence we need to show that the dual of any planar Eulerian graph is bipartite, and vice versa.

Let $G$ be an Eulerian planar graph. By definition, all vertices in $G$ are even in degree. Since the degree of a vertex in $G$ corresponds to the number of edges surrounding a face in the dual $G^*$, the edges surrounding each face in $G^*$ constitute cycles of even length. Hence according to Theorem 2.8, $G^*$ is bipartite.

Conversely, let $G^*$ be bipartite. This means $G^*$ contains no odd cycles and, since $G$ is planar, all faces are surrounded by an even number of edges. Hence all vertex degrees in $G$ are even, making $G$ Eulerian.                                        □

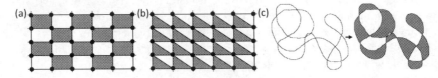

**Fig. 5.4** Examples of face colourings using two colours

Practical examples of Theorem 5.5 arise in the tiling industry where we are often interested in laying tiles of two different colours such that adjacent tiles do not have the same colour. Two example patterns are shown in Figure 5.4(a) and (b). Close examination of these patterns reveals the underlying graphs to be Eulerian as expected. Another example arises in the childhood doodling game in which a single connected line is drawn on a piece of paper, with the faces then being coloured using just two colours. Figure 5.4(c) shows an example of this game. We see that each time the line crosses itself, the degree of the vertex existing at this intersection increases by 2; hence vertex degrees are always even as expected.

The connection between face $k$-colourings of maps and vertex $k$-colourings of their planar duals allows us to conclude that the Four Colour Theorem for maps is equivalent to the statement that the vertices of all loop-free planar graphs are 4-colourable.[1] This concept was hinted upon in Section 1.2 where, in Figure 1.7, we took a map of Wales, constructed its (planar) dual graph, 4-coloured its vertices, and then converted this solution back into a 4-colouring of the faces of original map. The task of proving that four colours are sufficient for the vertices of *any* planar graph (and therefore the faces of any map) was formally one of the most famous unsolved problems in the whole of mathematics. It was eventually solved in controversial circumstances by Kenneth Appel and Wolfgang Haken in 1976. Their proof is both

---

[1] Recall that loops (i.e., edges of the form $\{v_i, v_i\}$) are disallowed in the vertex colouring problem.

very long and involved using months of computation time to test and classify a
large number of different graph configurations. Consequently, we end this section
by restricting ourselves to proving the weaker Six and Five Colour Theorems, before
giving a general history of the Four Colour Theorem itself.

**Theorem 5.6** *The vertices of any loop-free planar graph are 6-colourable.*

*Proof.* Let $G$ be a planar graph with $n \geq 3$. According to Theorem 5.2, $G$ has at
most $3n - 6$ edges. This means that the minimal degree of $G$ cannot exceed 5. Thus
in every subgraph $G'$ of $G$ there is a vertex with degree of at most $\delta = 5$. Therefore,
according to Theorem 2.6, we get $\chi(G) \leq 5 + 1$.                                      $\square$

With some additional reasoning we can improve this result to get the following:

**Theorem 5.7 (Heawood (1890))** *The vertices of any loop-free planar graph are 5-
colourable.*

*Proof.* For contradictory purposes, suppose this statement to be false, and let $G$ be a
planar graph with chromatic number $\chi(G) = 6$ and a minimal number of vertices $n$.
Because of Theorem 5.2, $G$ must have a vertex $v$ with $\deg(v) \leq 5$. Let $G' = G - \{v\}$.
We know that $G'$ can be 5-coloured using, say, colours labelled 1 to 5. Each of
these colours must also be used to colour at least one neighbour of $v$; otherwise
$G$ would also be 5-colourable. We can now assume that $v$ has five neighbours, say
$u_1, u_2, \ldots, u_5$, arranged in a clockwise fashion around $v$, with colours $c(u_i) = i$.

Now denote by $G'(i, j)$ the subgraph of $G'$ spanned by vertices with colours $i$ and
$j$. Suppose that $u_1$ and $u_3$ belong to separate components of $G'(1,3)$. Interchanging
the colours 1 and 3 in the component of $G'(1,3)$ containing $u_1$ will give us another
feasible 5-colouring of $G'$. However, in this 5-colouring, both $u_1$ and $u_3$ will have
the same colour, meaning that a spare colour now exists for $v$. This implies that $G$ is
in fact 5-colourable.

Since $u_1$ and $u_3$ must belong to the same component $G'(1,3)$, we now deduce
the existence of a path $P_{1,3}$ in $G'$ whose vertices are coloured using colours 1 and
3 only. Similarly, $G'$ must also contain a path $P_{2,4}$ using colours 2 and 4. However,
this is impossible in a planar graph since the cycle $u_1, P_{1,3}, u_3$ separates $u_2$ from $u_4$,
meaning that $P_{2,4}$ cannot be drawn without edges crossing (see Figure 5.5). Hence
$G$ cannot be planar.                                                                       $\square$

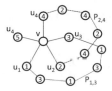

**Fig. 5.5** Depicting paths $P_{1,3}$ and $P_{2,4}$ used in the proof of Theorem 5.7. Colour labels are written
inside the vertices

## 5.1.2  Four Colours Suffice

In the proof of Theorem 5.7 we make use of the notation $G(i, j)$, which denotes the subgraph induced by taking only those vertices coloured with colours $i$ and $j$ in $G$. Note that individual components of $G(i, j)$ are actually Kempe chains (see Definition (3.1)). These are named after the mathematician Alfred Kempe (1849–1922), who used them in his infamous false proof for the Four Colour Theorem in 1879.

As we saw in Chapter 1, the conjecture that all maps can be coloured using at most four colours was first pointed out in 1852 by Francis Guthrie (1831–1899), who, at the time, was a student at University College London. Guthrie passed these observations on to his brother Frederick who, in turn, passed them on to his mathematics tutor Augustus De Morgan (1807–1871). De Morgan was not able to provide a conclusive proof for this conjecture, but the problem, being both easy to state and tantalisingly difficult to solve, captured the interest of many of the most notable mathematicians of the era, including William Hamilton (1805–1865), Arthur Cayley (1821–1895) and Charles Pierce (1839–1914). Indeed, over time the Four Colour Conjecture was to become one of the most famous unsolved problems in all of mathematics.

In 1879 a student of Arthur Cayley, Alfred Kempe, announced in *Nature* magazine that he had proved the Four Colour Theorem, publishing his result in the *American Journal of Mathematics* (Kempe, 1879). In his arguments, Kempe made use of his eponymous Kempe chains in the following way. Suppose we have a map in which all faces except one are coloured using colours 1, 2, 3, or 4. If the uncoloured face, which we shall call $F$, is not surrounded by faces featuring all four colours, then obviously we can colour $F$ using the missing colour. Therefore, suppose now that $F$ is surrounded by faces $F_1$, $F_2$, $F_3$, and $F_4$ (in that order), which are coloured using colours 1, 2, 3, and 4 respectively. There are now two cases to consider:

Case 1:  There exists no chain of adjacent faces from $F_1$ to $F_3$ that are alternately coloured with colours 1 and 3.

Case 2:  There *is* a chain of adjacent faces from $F_1$ to $F_3$ that are alternately coloured with colours 1 and 3.

If Case 1 holds then $F_1$ can be switched to colour 3, and any remaining faces in the chain can also have their colours interchanged. This operation retains the feasibility of the solution (no adjacent faces will have the same colour) and also means that no face adjacent to $F$ will have colour 1. Consequently $F$ can be assigned to this colour.

If Case 2 holds then there cannot exist a chain of faces from $F_2$ to $F_4$ using only colours 2 and 4. This is because, for such a chain to exist, it would need to cross the chain from $F_1$ to $F_3$, which is impossible on a map. Thus, Case 1 holds for $F_2$ and $F_4$, allowing us to switch colours as with Case 1.

The arguments of Kempe were widely accepted among mathematicians of the day; he was promptly elected a Fellow of the Royal Society and also went on to be knighted in 1912. The Four Colour *Conjecture* was now the Four Colour *Theorem*.

This all changed 11 years later when, in 1890, English mathematician Percy Heawood (1861–1955) shocked the mathematics fraternity by publishing an example map which exposed a flaw in Kempe's arguments (Heawood, 1890). Though he failed to supply his own proof, Heawood had shown that the Four Colour Theorem was indeed still a conjecture. In the same publication Heawood did show, however, that arguments analogous to Kempe's could be used to prove that all maps are 5-colourable, as we saw in Theorem (5.7). In later work, Heawood also proved that if the number of edges around each region of a map is divisible by 3 then the map can be 4-coloured.

As the decades passed, the problem that had first been pointed out by Guthrie in 1852 still remained unproven. Some piecemeal progress towards a solution was made with one proof showing that four colours were sufficient for colouring maps of up to 27 faces. This was followed by proofs for up to 31 faces, and then 35 faces. However, it would turn out that methods used by Kempe and his contemporaries in early papers would ultimately pave the way forward.

To start, the focus of research turned towards writing proofs concerning the vertices of loop-free planar graphs (i.e., the dual graphs of maps). In the first half of the twentieth century, researchers also concentrated their efforts on reducing these graphs to special cases that could be identified and classified. The idea was to produce a minimal set of configurations that could each be tested. Initially, this set was thought to contain nearly 9,000 members, which was considered far too large for mathematicians to study individually. This compelled some to turn towards using computers in order to design specialised algorithms for testing them.

Ultimately, the first conclusive proof of the Four Colour Theorem was produced in 1976 by mathematicians Kenneth Appel (1932-2013) and Wolfgang Haken (b. 1928), who showed that no configuration can exist that will appear in a minimal counterexample to the Four Colour Theorem (Appel and Haken, 1977a,b,c). In research carried out at the University of Illinois, together they reduced the set of configurations to just 1,936 members, which were then individually checked by computer. As was later stated in Appel's obituary in *The Economist* on May 4th, 2013:

> Both he and Dr Haken hugely exceeded their time allocation on the computer, which belonged to the university administration department. ... Their proof depended on both hand-checking by family members and then brute force computer power; the result was published in over 140 pages in the *Illinois Journal of Mathematics* and 400 pages of further diagrams on microfiche. They also, in the old fashioned way, chalked the message on a blackboard in the mathematics department: FOUR COLOURS SUFFICE.

At the time, this proof was controversial, with some mathematicians publicly questioning the legitimacy of a proof in which much of the work had been carried out by computer. (How might we guarantee the reliability of the algorithms and hardware?) However, despite these worries independent verification soon convinced the majority that the Four Colour Theorem had indeed finally been proved. Hence we are now able to state:

**Theorem 5.8 (The Four Colour Theorem)** *The vertices of any loop-free planar graph are 4-colourable. Equivalently, the faces of any map are 4-colourable.*

In recent years, refinements to Haken and Appel's proof have been made, with Robertson et al. (1997) showing that, through various acts of trickery, only 633 configurations need to be considered. (A short summary of this proof, together with a description of a polynomial-time algorithm for 4-colouring planar graphs, can be found at http://people.math.gatech.edu/~thomas/FC/fourcolor.html.) However, a proof along more "traditional" lines still remains elusive and, to this day, the Four Colour Theorem remains an excellent example, along with Fermat's Last Theorem, of a problem that is very easy to state, but exceptionally difficult to prove.

Readers interested in finding out more about the fascinating history of the Four Colour Theorem are invited to consult the very accessible book *Four Colors Suffice: How the Map Problem Was Solved* by Wilson (2003).

## 5.2 Edge Colouring

Another way in which graphs might be coloured is by assigning colours to their *edges*, as opposed to their vertices or faces. This gives rise to the *edge colouring problem* where we seek to colour all edges of a graph so that no pair of edges sharing a common vertex (incident edges) have the same colour, and so that the number of colours used is minimal. The edge colouring problem has applications in scheduling round robin tournaments and also transferring files in computer networks (de Werra, 1988; Coffman et al., 1985). The minimum number of colours needed to edge colour a graph $G$ is called the *chromatic index*, denoted by $\chi'(G)$. This should not be confused with the *chromatic number* $\chi(G)$, which is the minimum number of colours needed to *vertex* colour a graph $G$.

As mentioned earlier, the edge colouring and vertex colouring problems are very closely related because we are able to edge colour any graph by simply vertex colouring its corresponding *line graph*.

**Definition 5.3** *Given a graph G, the* line *graph of G, denoted by L(G), is constructed by using each edge in G as a vertex in L(G), and then connecting pairs of vertices in L(G) if and only if the corresponding edges in G share a common vertex as an endpoint.*

An example conversion between a graph $G$ and its line graph $L(G)$ is shown in Figure 5.6(a). From this process, it is natural that the number of vertices and edges in $L(G)$ is related to the number of vertices and edges in $G$.

**Theorem 5.9** *Let $G = (V, E)$ be a graph with n vertices and m edges. Then its line graph $L(G)$ has m vertices and*

$$\frac{1}{2} \sum_{v \in V} \deg(v)^2 - m$$

*edges.*

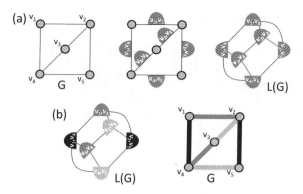

**Fig. 5.6** Illustration of (a) how to convert a graph $G$ into its line graph $L(G)$, and (b) how a vertex $k$-colouring of $L(G)$ corresponds to an edge $k$-colouring of $G$

*Proof.* Since each edge in $G$ corresponds to a vertex in $L(G)$ it is obvious that $L(G)$ has $m$ vertices. Now let $\{u,v\}$ be an edge in $G$. This means that $\{u,v\}$ is a vertex in $L(G)$ with degree $\deg(u) + \deg(v) - 2$. Hence the total number of edges in $L(G)$ is:

$$\frac{1}{2} \sum_{\{u,v\}\in E} (\deg(u) + \deg(v) - 2) = \frac{1}{2} \sum_{\{u,v\}\in E} (\deg(u) + \deg(v)) - m$$

Note that the degree of each vertex $v$ appears exactly $\deg(v)$ times in this sum. Hence, we can simplify the expression to that stated in Theorem 5.9 as required.

□

Figure 5.6(b) also demonstrates the way in which a vertex $k$-colouring of the line graph $L(G)$ corresponds to an edge colouring of $G$. Consequently, rather like the way in which a face colouring problem can be solved by vertex colouring a graph's dual, any edge colouring problem stated on a graph $G$ can be tackled by applying a vertex colouring algorithm to its line graph $L(G)$.

We now discuss some important results concerning the chromatic index of a graph.

**Theorem 5.10** *Let $K_n$ be the complete graph with $n > 1$ vertices. Then $\chi'(K_n) = n - 1$ if $n$ is even; else $\chi'(K_n) = n$.*

*Proof.* When $n$ is odd, the edges of $K_n$ can be coloured using $n$ colours by the following process. First, draw the vertices of $K_n$ in the form of a regular $n$-sided polygon. Next, select an arbitrary edge on the boundary of this polygon and colour it, together with all edges parallel to it, using colour 1. Now moving in a clockwise direction, select the next edge on the boundary and colour it, together with its parallel edges, with colour 2. Continue this process until all edges have been coloured.

It is easy to demonstrate that the edges of $K_n$ are not $(n-1)$-colourable by the fact that the largest number of edges that can be assigned the same colour is $(n-1)/2$;

it then follows, because the number of edges in $K_n$ is $\frac{n(n-1)}{2}$, that $n$ colours are required.

When $n$ is even, a similar process can be followed, where a regular $(n-1)$-sided polygon is constructed, with the remaining vertex being placed in the centre. The same method for the $(n-1)$ case is then followed, with edges perpendicular to the edges currently being coloured also being assigned to the same colour. As in the previous case it is easily shown that no feasible edge colouring of $K_n$ exists using fewer than $n$ colours.                                                                    □

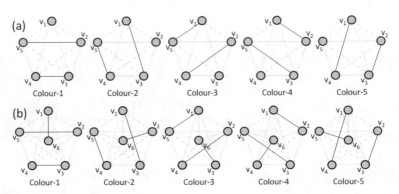

**Fig. 5.7** Illustrating how optimal edge colourings can be constructed for complete graphs with (a) $K_5$ and (b) $K_6$ using the circle method

The method used in the proof of Theorem 5.10 is often referred to as the circle (or polygon) method and was originally proposed by mathematician and Church of England minister Thomas Kirkman (1806-1895) (Kirkman, 1847). An important practical use of this method is for constructing round-robin sports leagues, where we have a set of $n$ teams that are required to play each other once across a sequence of rounds. Figure 5.7 provides examples of this method for $n = 5$ and $n = 6$. Here, the vertices can be thought of as "teams", with edges representing "matches" between these teams. Each colour then represents a round in the schedule. Considering Figure 5.7(a), where $n = 5$, the first round involves matches between team-$v_2$ and team-$v_5$ and between team-$v_3$ and team-$v_4$, with team-$v_1$ receiving a bye. The next round then involves matches between team-$v_1$ and team-$v_3$ and between team-$v_4$ and team-$v_5$, with team-$v_2$ receiving a bye, and so on. The pattern is similar when $n$ is even, as shown in Figure 5.7(b), except that no team receives a bye. Applications of graph colouring to sports scheduling problems are considered in more detail in Chapter 7.

A further result, due to König (1916), concerns the chromatic index of bipartite graphs.

**Theorem 5.11 (Konig's Line Colouring Theorem)** *Let $G = (V_1, V_2, E)$ be a bipartite graph with maximal degree $\Delta(G)$. Then $\chi'(G) = \Delta(G)$.*

*Proof.* The proof is via induction on the number of edges $m$ in $G$. It is sufficient to prove that if $m-1$ edges have been coloured using at most $\Delta(G)$ colours, then the remaining edge can be coloured using one of the $\Delta(G)$ available colours.

Suppose all edges except $\{u,v\}$ have been coloured. Then there exists at least one colour not incident to $u$, and one colour not incident to $v$. If the same colour is not incident to both $u$ and $v$, then edge $\{u,v\}$ can be assigned to this colour. If this is not the case, without loss of generality we can say that: (a) $u \in V_1$ is incident to a grey edge, but not a black edge; and (b) $v \in V_2$ is incident to a black edge but not a grey edge.

Now consider a grey-black Kempe chain starting from $u$ (that is, a component of $G$ containing $u$ that comprises only those vertices incident to a grey or black edge). Travelling along this chain from $u$ involves alternating between vertex sets $V_1$ and $V_2$. However, we will never reach $v$ because each time we arrive at a vertex in $V_2$ we do so via a grey edge, which $v$ is not incident to.

Since $v$ is not in the Kempe chain, we can now interchange the colours of the edges in this chain without affecting $v$ (or indeed any other vertices outside of the chain). Hence the edge $\{u,v\}$ can be coloured grey, completing the edge colouring.

$\square$

The previous two theorems demonstrate that the edge colouring problem is solvable in polynomial time for both complete and bipartite graphs. We have also seen that, for both topologies, their chromatic indices $\chi'(G)$ are either $\Delta(G)$ or $\Delta(G)+1$. Somewhat surprisingly, it also turns that this feature applies to *any* graph $G$, as proved by Vizing (1964).

**Theorem 5.12 (Vizing's Theorem)** *Let $G$ be a simple graph with maximal degree* $\Delta(G)$; *then* $\Delta(G) \leq \chi'(G) \leq \Delta(G)+1$.

*Proof.* When $\Delta(G)$ edges are incident to a vertex, these edges all require a different colour. Hence the lower bound is proved: $\Delta(G) \leq \chi'(G)$.

The upper bound can be proved via induction on the number of edges. Suppose that, using $\Delta(G)+1$ colours, we have coloured all edges in $G$ except for the single edge $\{u,v_0\}$. Since $\Delta(G)$ gives the maximal degree, at least one colour will be unused at each of these two vertices. Now construct a series of edges, $\{u,v_0\}, \{u,v_1\}, \ldots$, and a sequence of colours, $c_0, c_1, \ldots$, as follows: Select a colour $c_i$ that is an unused colour at $v_i$. Now, let $\{u,v_{i+1}\}$ be an edge with colour $c_i$. We stop (with $i=k$) when either $c_k$ is an unused colour at $u$, or $c_k$ is already used on an edge $\{u,v_{j<k}\}$.

Case 1: If $c_k$ is an unused colour at $u$, then we can recolour $\{u,v_i\}$ with $c_i$ for $0 \leq i \leq k$. We now need to simply recolour edges incident to $u$ to complete the proof.

Case 2: Otherwise, we recolour $\{u,v_i\}$ with $c_i$ for $0 \leq i < j$ and remove the colour from $\{u,v_j\}$. Observe that $c_k$ (say, "grey") is missing at both $v_j$ and $v_k$. Now let "black" be an used colour at $u$. If grey is unused at $u$ then we can colour $\{u,v_j\}$ grey. If black is unused at $v_j$ then we can colour $\{u,v_j\}$ black. However, if black is unused at $v_k$ then we colour $\{u,v_i\}$ with $c_i$ for $j \leq i < k$ and colour $\{u,v_k\}$

black, because none of the edges $\{u, v_i\}$ for $j \leq i < k$ will be coloured grey or black.

If neither case above holds, then we consider the subgraph of grey and black edges. The components of this subgraph will be paths and/or cycles. The vertices $u$, $v_i$, and $v_k$ are the terminal vertices of paths; hence they cannot all belong to the same component. In this case, select a component containing just one of these vertices and interchange the colours of its edges. This means that one of the cases above now applies. □

In essence, Vizing's theorem tells us that the set of all graphs can be partitioned into two classes: "class one" graphs, for which $\chi'(G) = \Delta(G)$, and "class two" graphs, where $\chi'(G) = \Delta(G) + 1$. Holyer (1981) has shown that the problem of testing whether a graph belongs to class one is NP-complete. However, a number of polynomially bounded algorithms are available for colouring the edges of any graph using exactly $\Delta(G) + 1$ colours, such as the $\mathcal{O}(nm)$ algorithm of Misra and Gries (1992). The existence of such algorithms tells us that we can colour the edges of any graph using a maximum of one extra colour beyond its chromatic index.

We might now ask whether the existence of such tight bounds for the edge colouring problem helps us to garner further information about the vertex colouring problem. It is clear that if we were given the task of vertex colouring a line graph $L(G)$, one approach would be to convert $L(G)$ into its "original" graph $G$, and then solve the corresponding edge colouring problem on $G$. Since $\chi(L(G)) = \chi'(G)$, then according to Vizing's theorem this would immediately tell us that we need to use either $\Delta(G)$ or $\Delta(G) + 1$ colours to feasibly colour the vertices of $L(G)$. Indeed, if $G$ were a type two graph, then algorithms such as Mistra and Gries's could be used to quickly find the optimal edge colouring for $G$ and therefore the optimal vertex colouring for $L(G)$. However, it should be remembered that this very attractive sounding proposal is only applicable when we wish to colour the vertices of a line graph that therefore has an "original" graph into which it can be converted. Unfortunately we cannot convert all graphs into an "original" graph in this way.

## 5.3 Precolouring

In the precolouring problem we are given a graph $G$ for which some subset of the vertices $V' \subseteq V$ has already been assigned to colours. Our task is to then colour the remaining vertices in the set $V - V'$ so that the resultant solution is feasible and uses a minimal number of colours.

Applications of precolouring arise in register allocation problems (see Section 1.1.4) where certain variables must be assigned to specific registers, perhaps due to calling conventions or communication between modules. They also occur in areas such as timetabling and sports scheduling where we might be given a problem instance in which some of the events have been preassigned to particular timeslots.

Precolouring problems can be easily converted into a standard graph colouring problem using graph contraction operations.

**Definition 5.4** *The* contraction *of a pair of vertices* $v_i, v_j \in V$ *in a graph G produces a new graph in which* $v_i$ *and* $v_j$ *are removed and replaced by a single vertex v such that v is adjacent to the union of the neighbourhoods of* $v_i$ *and* $v_j$*; that is,* $\Gamma(v) = \Gamma(v_i) \cup \Gamma(v_j)$.

The following steps can now be taken. Given a precolouring problem instance defined on a graph $G$, let $V'(i)$ define the set of vertices precoloured with colour $i$. Assuming there are $k$ different colours used in the precolouring, $\bigcup_{i=1}^{k} V'(i) = V'$ and $V'(i) \cap V'(j) = \emptyset$, for $1 \le i \ne j \le k$. First, for each set $V'(i)$, merge all vertices into a single vertex using a series of contraction operations. This has the effect of reducing the number of precoloured vertices to $k$. Next, add edges between each pair of the $k$ contracted vertices to form a clique. Finally remove all colours from the vertices of this graph, and apply any arbitrary graph colouring algorithm to produce a feasible solution. A colouring of the original can then be obtained by simply reversing the above process. An example is provided in Figure 5.8.

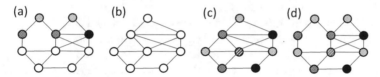

**Fig. 5.8** How a precolouring problem (a), can be converted into a new graph via contraction operations on the precoloured vertices (b), then coloured (c), and then converted back into a solution to the original problem (d)

## 5.4  Latin Squares and Sudoku Puzzles

Another prominent area of mathematics for which graph colouring techniques are naturally suited is the field of Latin squares. Latin squares are $l \times l$ grids that are filled with $l$ different symbols, each occurring exactly once per row and once per column. They were originally considered in detail by Leonhard Euler, who filled his grids with symbols from the Latin alphabet, though nowadays it is common to use the integers 1 through to $l$ to fill the grids. Example Latin squares of different sizes are shown in Figure 5.9.

Latin squares have practical applications in various different areas, including scheduling and experimental design. For an application in scheduling, imagine that we have two groups of $l$ people and we want to schedule meetings between all pairs of people belonging to different groups. Clearly in this case, $l^2$ meetings will take place in total and, since only $l$ meetings can take place simultaneously, at least $l$

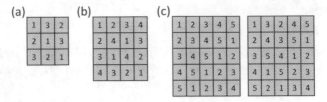

**Fig. 5.9** Example Latin squares for (a) $l = 3$, (b) $l = 4$, and (c) $l = 5$

timeslots will be needed. Latin squares give solutions to such problems that make use of exactly $l$ timeslots. To see this, let us name the members of team one as $r_1, r_2, \ldots, r_l$, which are represented by the rows in the grid, and the members of team two $c_1, c_2, \ldots, c_l$, represented by the columns. The characters within an $l \times l$ Latin square then represent the various timeslots to which the meetings are assigned. For example, the Latin square shown in Figure 5.9(a) schedules meetings between $r_1$ and $c_1$, $r_2$ and $c_2$, and $r_3$ and $c_3$ into timeslot 1; meetings between $r_1$ and $c_3$, $r_2$ and $c_1$, and $r_3$ and $c_2$ into timeslot 2; and meetings between $r_1$ and $c_2$, $r_2$ and $c_3$, and $r_3$ and $c_1$ into timeslot 3. Obviously, any $l \times l$ Latin square will provide a suitable meeting schedule fitting these criteria.

For an example application of Latin squares in experimental design, imagine that a medical trials centre wants to test the effects of $l$ different drugs on a particular illness. Suppose further that the trials are to take place over $l$ weeks using $l$ different patients, with each patient receiving a single drug in each week. An $l \times l$ Latin square can be used to allocate treatments in this case, with rows representing patients, and columns representing weeks. This means that over the course of the $l$ weeks each patient receives each of the $l$ drugs once, and in each week all of the $l$ drugs are tested. Looking at the $3 \times 3$ Latin square from Figure 5.9(a), for example, we see that Patients 1, 2, and 3 are administered Drugs 1, 2, and 3 (respectively) in Week 1; Drugs 3, 1, and 2 in Week 2; and Drugs 2, 3, and 1 in Week 3, as required.

Note that we are able to permute the rows and columns of a Latin square and still retain the property of each character occurring exactly one per column and once per row. It is therefore common to write Latin squares in their *standardised form*, whereby the rows and columns are arranged so that the top row and leftmost column of the grid have the characters in their natural order $1, 2, \ldots, l$. The other $l!(l-1)! - 1$ Latin squares that can be formed by permuting the rows and columns are then considered to be equivalent to this. The Latin square in Figure 5.9(b) is in standardised form, while the one in Figure 5.9(a) is not. It is also known that as $l$ is increased, then so does the number of different $l \times l$ Latin squares. For $l = 11$ this figure is approximately $5.36 \times 10^{33}$, though for larger values of $l$ these figures have so far proved too large to compute.

Figure 5.10 shows how the production of a Latin square can be expressed as a graph colouring problem. Here, as illustrated, the symbols used within the grid represent the colours. Each cell of the grid is then associated with a vertex, and edges are added between all pairs of vertices in the same row, and all pairs of vertices in the same column. This results in a graph $G = (V, E)$ with $n = l^2$ vertices and

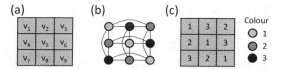

**Fig. 5.10** Showing the relationship between graph colouring and Latin squares. Part (a) associates each grid cell with a vertex; (b) shows the corresponding graph together with a feasible colouring; and (c) gives a valid Latin square corresponding to this colouring

$m = l^2(l-1)$ edges, for which $\deg(v) = 2(l-1) \; \forall v \in V$. We also see that the set of vertices in each row forms a clique of size $l$, as do vertices in each column, implying that solutions using fewer than $l$ colours are not possible.

Note that it is simple to produce a Latin square in standardised form for any value of $l$ by simply using values $(1, 2, \ldots, l)$ for the first row, $(2, 3, \ldots, l, 1)$ for the second row, $(3, 4, \ldots, l, 1, 2)$ for the third row, and so on (see for example the left Latin square in Figure 5.9(c)). Hence graphs representing Latin squares are actually a particular topology for which the associated graph colouring problem can be easily solved in polynomial time for any value of $l$, without a need for resorting to heuristics or approximation algorithms. Graph colouring algorithms can, however, be used for producing different Latin squares to this.

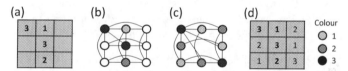

**Fig. 5.11** A partial $3 \times 3$ Latin square with four filled-in cells (a); the corresponding precolouring problem (b); the same graph with a contraction of two vertices, together with a feasible colouring (c); and the corresponding Latin square solution (d)

Graph colouring algorithms arguably become more useful in this area when we consider the *partial* Latin square problem. This is the problem of taking a partially filled $l \times l$ grid and deciding whether or not it can be completed to form a Latin square. This problem has been shown to be NP-complete by Colbourn (1984).

Figure 5.11 gives an example of how the partial Latin square problem can be tackled using graph colouring principles. In essence it follows the same method as the previous example given in Figure 5.10, except that extra edges can now also be added between any pair of vertices predefined as having the same colour. Once this has been done, the same steps as with the precolouring problem (Section 5.3) can be followed, with contractions being used to make the graph smaller if desired. An $l$-colouring of this graph then corresponds to a completed $l \times l$ Latin square. Of course, depending on the values of the filled-in cells in the original problem, there could be zero, one, or multiple feasible $l$-colourings available.

## 5.4.1 Solving Sudoku Puzzles

The partial Latin square problem has become very popular in recent decades in the form of Sudoku puzzles. In Sudoku we are given a partially filled Latin square and the objective is to complete the remaining cells so that each column and row contains the characters $1, \ldots, l$ exactly once. In addition, Sudoku grids are also divided into $l$ "boxes" (usually marked by bold lines) which are also required to contain the characters $1, \ldots, l$ exactly once; thus Sudoku can be considered a special case of the partial Latin square problem in which the constraint of appropriately filling out the "boxes" must also be satisfied. An example $9 \times 9$ Sudoku puzzle and corresponding solution is shown in Figure 5.12.

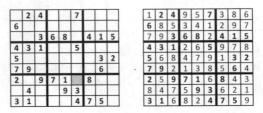

**Fig. 5.12** A $9 \times 9$ Sudoku puzzle and corresponding solution

Because Sudoku is intended to be an enjoyable puzzle, problems posed in books and newspapers will nearly always be *logic solvable*.

**Definition 5.5** *A Sudoku puzzle is* logic solvable *if and only if it features exactly one solution, which is achievable via forward chaining logic only.*

Puzzles that are not logic solvable require random choices to be made. In general these should be avoided because players will have to go through the tedious process of backtracking and reguessing if their original guesses turn out to be wrong.

As an example of how a player might deduce the contents of cells, consider the puzzle given in Figure 5.12. Here, we see that the cell in the seventh row and sixth column (shaded) must contain a 6, because all numbers 1 to 5 and 7 to 9 appear either in the same column, the same row, or the same box as this cell. If the problem instance is logic solvable (as indeed this one is), the filling in of this cell will present further clues, allowing the user to eventually complete the puzzle.

A number of algorithms for solving Sudoku puzzles are available online, such as those at www.sudokuwiki.org and www.sudoku-solutions.com. Such algorithms typically mimic the logical processes that a human might follow, with popular deductive techniques, such as the so-called X-wing and Swordfish rules, also being commonplace. In other areas of Sudoku research, Russell and Jarvis (2005) have shown that the number of essentially different Sudoku solutions (when symmetries such as rotation, reflection, permutation and relabelling are taken into account) is 5,472,730,538 for the popular $9 \times 9$ grids. McGuire et al. (2012) have also shown that $9 \times 9$ Sudoku puzzles must contain at least 17 filled-in cells to be logic solvable (thus $9 \times 9$ puzzles with 16 or fewer filled-in cells will always admit more than one solution). Similar results for larger grids are unknown, however. Herzberg and Murty (2007) have also shown that at least $l - 1$ of the $l$ characters must be present in the filled cells of a Sudoku puzzle for it to be logic solvable.

Despite the fact that Sudoku is a special case of the partial Latin squares problem, Yato and Seta (2003) have demonstrated that the problem of deciding whether or not a Sudoku puzzle features a valid solution is still NP-complete. Graph colouring algorithms can therefore be useful for solving instances of Sudoku, particularly those that are not necessarily logic solvable.

Sudoku puzzles can be transformed into a corresponding graph colouring problem in the same fashion as partial Latin square problems (see Figure 5.11), with additional edges also being imposed to enforce the extra constraint concerning the "boxes" of the grid. We now present two sets of experiments that illustrate the capabilities of the HEA and backtracking algorithms from Chapter 4 for solving Sudoku puzzles. In the first set of experiments we focus on Sudoku problems that are not necessarily logic solvable (random puzzles), while in the second set we focus on $9 \times 9$ grids that are logic solvable.

### 5.4.1.1 Solving Random Sudoku Puzzles

To generate problem instances that are not necessarily logic solvable, we can start by taking a completed Sudoku solution of a given size. Such solutions can be obtained from a variety of places such as the solution pages of a Sudoku book or newspaper, or by simply executing a suitable graph colouring algorithm on a graph representing a blank puzzle. In the next step of the procedure, this completed grid can then be randomly shuffled using the following five operators:

- Transpose the grid;
- Permute columns of boxes within the grid;
- Permute rows of boxes within the grid;
- Permute columns of cells within columns of boxes; and
- Permute rows of cells within rows of boxes.

Note that all of these shuffle operators preserve the validity of a Sudoku solution. Finally, a number of cells in the grid are made blank by going through each cell in turn and deleting its contents with probability $1 - p$, where $p$ is a parameter to be defined by the user. This means that instances generated with a low value for $p$ have a lower proportion of filled-in cells.

Figure 5.13 illustrates the performance of the HEA and backtracking algorithms on $9 \times 9$, $16 \times 16$ and $25 \times 25$ Sudoku grids respectively. In each case 100 instances for each value of $p$ have been generated and, as in Chapter 4, a computation limit of $5 \times 10^{11}$ constraint checks is imposed. For each algorithm two statistics are displayed. The *success rate* (SR) indicates the percentage of runs for which the algorithms have found a valid Sudoku solution (a feasible $l$-colouring) within the computation limit. The *solution time* then indicates the mean number of constraint checks that it took to achieve these solutions. Note that only successful runs are considered in the latter statistic.

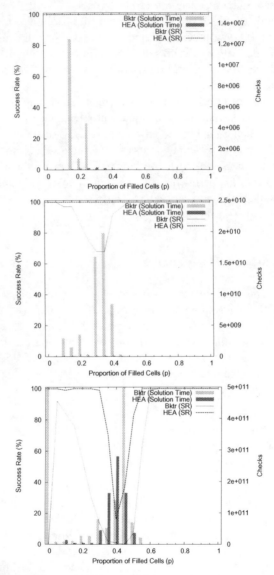

**Fig. 5.13** Comparison of the HEA and backtracking algorithm's performance with random Sudoku instances of size $9 \times 9$, $16 \times 16$, and $25 \times 25$ respectively. Note the different scales on the vertical axes in each case

Looking at the results for $9 \times 9$ Sudoku puzzles first, we see that both algorithms feature a 100% success rate across all instances with only a very small proportion of the computation limit being required.[2] For $16 \times 16$ puzzles, similar patterns occur for the HEA, with all problem instances being solved, and no runs requiring more than one second of computation time. On the other hand, the backtracking algorithm features a dip in its success rate for values of $p$ between 0.1 to 0.55, with a corresponding increase in solution times. Finally, with the larger $25 \times 25$ puzzles, this pattern becomes even more apparent, with both algorithms featuring dips in their success rates and subsequent increases in their solution times. However, these dips are less pronounced with the HEA, indicating its superior performance overall.

The dips in the success rates of these algorithms are analogous to the phase transition regions we saw with the flat graphs in Section 4.2. When $p$ is low, although solution spaces will be larger, there will also tend to be many valid solutions within these spaces. Consequently, an effective algorithm should be able to find one of these within a reasonable amount of computation time, as is the case with the HEA. For high vales for $p$ meanwhile, although there will only be a very small number of valid solutions (and perhaps only one), the solution space will be much smaller. Additionally, solutions to these highly constrained instances will also tend to reside at prominent optima (with a strong basin of attraction), thus also allowing easy discovery by an effective algorithm. However, instances at the boundary of these two extremes will cause greater difficulties. First, the solution spaces for these instances will still be relatively large, but they will also tend to admit only a small number of valid solutions. Second, because of their moderate number of constraints, the cost landscapes will also tend to feature more plateaus and local optima, making navigation towards a global optimum more difficult for the algorithm.

### 5.4.1.2 Logic Solvable Sudoku Puzzles

We now examine the performance of the HEA and backtracking algorithms on logic solvable instances, allowing us to examine the effect that the size of the solution space has on the difficulty of a puzzle when its solution is known to be unique. As we have seen, it is known that a $9 \times 9$ Sudoku puzzle must contain at least 17 filled-in cells to be logic solvable (McGuire et al., 2012). There is also an online resource containing over 49,000 different instances of these 17-clue puzzles maintained by Gordon Royle, available at:

staffhome.ecm.uwa.edu.au/~00013890/sudokumin.php

For our tests we took a random sample of 100 of these 17-clue instances together with their corresponding (unique) solutions. Logic solvable puzzles with more than 17 filled-in cells were then also generated for each of these by randomly selecting an appropriate number of blank cells in the puzzles, and adding the corresponding

---

[2] On our equipment (3.0 GHz Windows 7 PC with 3.87 GB RAM) the longest run in the entire set took just 0.02 seconds.

entries from their solutions. This operation maintains the uniqueness of each puzzle's solution while reducing the size of its solution space. All other experimental details remain the same as in the previous subsection.

Figure 5.14 shows the relative performance of the two graph colouring algorithms. In all cases, valid solutions were found within the computation limit. When the solution space size is relatively large (17-20 filled-in cells) we see that the HEA requires up to $2.4 \times 10^{10}$ constraint checks to find a solution.[3] However, beyond this point the puzzles are solved using very little computational effort—indeed, for more than 35 filled-in cells, solutions are achieved by the initial solutions produced by the DSATUR algorithm.

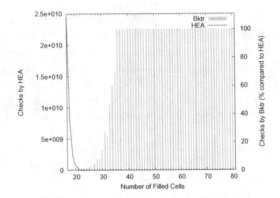

**Fig. 5.14** Comparison of the HEA and backtracking algorithm's performance on $9 \times 9$ Sudoku grids with unique solutions

In contrast to our earlier results on random Sudoku puzzles, Figure 5.14 also shows that the backtracking algorithm outperforms the HEA with these problem instances. With 17-clue puzzles for example, the algorithm has identified the unique solutions using just 0.03% of the computational effort required by the HEA. Thus, unlike in the larger puzzles seen in the previous section, here the solution spaces seem suitably sized and structured for the backtracking algorithm to be able to identify the unique Sudoku solutions in very short spaces of time.

## 5.5 Short Circuit Testing

Another interesting practical application of graph colouring is due to Garey et al. (1976), who suggest its use in the process of testing for (undesired) short circuits in printed circuit boards. In their model, a circuit board is represented by a finite lattice of evenly spaced points on to which a set of $n$ cycle-free components has

---

[3] This equated to approximately three minutes on our computer.

been printed. This set $\mathcal{P}$ is referred to as a *net pattern*, with individual components $P \in \mathcal{P}$ being called *nets*. Each net connects a number of points that are intended to be electrically common. An example net pattern comprising four components is shown in Figure 5.15(a). Note that connections between points are only permitted in vertical or horizontal directions.

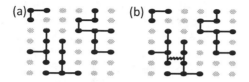

**Fig. 5.15** An example net pattern (a), and a net pattern containing a short between two nets (b)

Given a net pattern $\mathcal{P}$, the problem of interest is to determine whether there exists some fault on the circuit board (due to the manufacturing process) whereby an extra conductor path has been introduced between two nets that are not intended to be electrically common. This is the case in Figure 5.15(b). These extra conductor paths are known as "shorts".

An obvious strategy to determine whether a short has occurred is to test each pair of nets $P_i, P_j \in \mathcal{P}$ in turn by applying an electrical current to $P_i$ and seeing if this current spreads to $P_j$. However, Garey et al. (1976) suggest that the number of pairwise tests can be reduced significantly by making use of the following two observations.

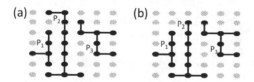

**Fig. 5.16** Two further net patterns

First, it is noted that many pairs do not need to be tested. Consider, for example, the net pattern in Figure 5.16(a). Here it is unnecessary to test the pair $P_1, P_3$ because if there is a short between them, then shorts must also exist between pairs $P_1, P_2$ and $P_2, P_3$. Since the objective of the problem is to determine if *any* shorts exist, testing either $P_1, P_2$ or $P_2, P_3$ is therefore sufficient. Furthermore, if we go on to consider the net pattern in Figure 5.16(b), it might also be reasonable to assume that shorts cannot occur between $P_1$ and $P_3$ without also causing a short involving $P_2$. Thus, depending on the criteria used for deciding where and how shorts can occur, we have the opportunity to exclude many pairs of nets from the testing procedure. If it is deemed necessary to test a pair of nets, these are called *critical pairs*; otherwise they are deemed *noncritical*.

The second observation is as follows. Let $G = (V,E)$ be a graph with a set of vertices $V = \{v_1, v_2, \ldots, v_n\}$, where each vertex $v_i \in V$ corresponds to a particular net $P_i \in \mathcal{P}$ (for $1 \leq i \leq n$). Also, let each edge $\{v_i, v_j\} \in E$ correspond to a pair of nets $P_i, P_j$ judged to be critical. Now let $\mathcal{S} = \{S_1, \ldots, S_k\}$ be a partition of $V$ such that no pair of vertices $v_i, v_j$ in any subset $S_l$ forms a critical pair. From a graph colouring perspective, $\mathcal{S}$ therefore defines a feasible $k$-colouring of the vertices of $G$. Now suppose that for the printed circuit board in question, external conductor paths are provided so that all nets in any subset $S_l$ can be made electrically common during testing. There are thus $k$ "supernets" that need to be tested. It can be seen that the printed circuit board contains no short if and only if no pair of "supernets" is seen to be electrically common. Therefore, we only have to perform a maximum of $\binom{k}{2}$ tests as opposed to our original figure of $\binom{n}{2}$ tests. Naturally, it is desirable to reduce $k$ as far as possible in order to minimise the number of tests needed.

In their paper, Garey et al. (1976) propose a number of criteria for deciding whether a pair of nets should be deemed critical, with associated theorems then being presented. We now review some of these.

**Theorem 5.13 (Garey et al. (1976))** *Consider a pair of nets $P_i, P_j \in \mathcal{P}$ to be critical if and only if a straight vertical line of sight can be drawn that connects $P_i$ and $P_j$. Then the corresponding graph $G = (V,E)$ is planar and has a chromatic number $\chi(G) \leq 4$.*

*Proof.* Given a net pattern $\mathcal{P}$, for each pair of nets for which a vertical line of sight exists, draw such a line. Since each line is vertical, none can intersect. It is now possible to contract each net into a single point, deforming the lines of sight (which may no longer be straight lines) in such a way that they remain nonintersecting. This structure now corresponds to the graph $G = (V,E)$, with each vertex corresponding to a contracted net, and each edge corresponding to the lines of sight. Since $G$ is planar, $\chi(G) \leq 4$ according to the Four Colour Theorem (Theorem 5.8).          □

**Theorem 5.14 (Garey et al. (1976))** *Consider a pair of nets $P_i, P_j \in \mathcal{P}$ to be critical if and only if a straight vertical line of sight or a straight horizontal line of sight can be drawn that connects $P_i$ and $P_j$. Then the corresponding graph $G = (V,E)$ has a chromatic number $\chi(G) \leq 12$.*

*Proof.* It is first necessary to show that any graph $G$ formed in this way has a vertex $v$ with $\deg(v) \leq 11$. Let $G_1 = (V, E_1)$ and $G_2 = (V, E_2)$ be subgraphs of $G$ such that $E_1$ is the set of edges formed from vertical lines and $E_2$ is the set of edges formed from horizontal lines. Hence $E = E_1 \cup E_2$. By Theorem (5.13), both $G_1$ and $G_2$ are planar. We can also assume without loss of generality that the number of vertices $n > 12$. According to Theorem (5.2), the number of edges in a planar graph with $n$ vertices is less than or equal to $3n - 6$. Thus:

$$m \leq |E_1| + |E_2| \leq (3n - 6) + (3n - 6) = 6n - 12.$$

Since each edge contributes to the degree of two distinct vertices, this gives:

$$\sum_{v \in V} \deg(v) = 2m \leq 12n - 24.$$

Hence it follows that some vertex in $G$ must have a degree of 11 or less.

Now consider any subset $V' \subseteq V$ with $|V'| > 12$. The induced subgraph of $V'$ must contain a vertex with a degree at most 11. Consequently, according to Theorem (2.6), $\chi(G) \leq 11 + 1$.                                                                    $\square$

In their paper, Garey et al. (1976) conjecture that the result of Theorem (5.14) might actually be improved to $\chi(G) \leq 8$ because, in their experiments, they were not able to produce graphs featuring chromatic numbers higher than this. They also go on to consider the maximum length of lines of sight and show that:

- If lines of sight can be both horizontal and vertical but are limited to a maximum length of 1 (where one unit of length corresponds to the distance between a pair of vertically adjacent points or a pair of horizontally adjacent points on the circuit board), then $G$ will be planar, giving $\chi(G) \leq 4$.
- If lines of sight can be both horizontal and vertical but are limited to a maximum length of 2, then $G$ will have a chromatic number $\chi(G) \leq 8$.

Finally, they also note that if arbitrarily long lines of sight travelling in *any* direction are permitted (as opposed to merely horizontal or vertical) then it is possible to form all sorts of different graphs, including complete graphs. Hence arbitrarily high chromatic numbers can occur.

## 5.6  Graph Colouring with Incomplete Information

In this section we now consider graph colouring problems for which information about a graph is incomplete at the beginning of execution. In the following subsections we discuss three different interpretations, specifically decentralised graph colouring, online graph colouring, and dynamic graph colouring, and give practical examples of each.

### 5.6.1 Decentralised Graph Colouring

In decentralised graph colouring it is generally assumed that each vertex of a graph is an independent entity responsible for choosing its own colour. Moreover, the only information available to each vertex is who its neighbours are, and what their colours are. In other words, vertices are unaware of the structure of the graph beyond their own neighbouring vertices.

A practical example of this problem might occur in the setting-up of a wireless ad hoc network. Imagine a situation where a network is to be created by randomly dropping a set of wireless devices (equipped with radio transmitters and receivers) into a particular environment. Imagine further that each device in this network will broadcast information at a particular frequency on the radio spectrum. Finally, also consider the fact that if two devices are close together but using the same frequency (or frequencies that are suitably similar) their transmissions will interfere with one another, inhibiting the ability of other devices to decipher their individual signals.

The above situation is illustrated in Figure 5.17, where each wireless device appears at the centre of a grey circle denoting its transmission range. Figure 5.17(a) shows two devices, $u$ and $v$, that are situated in each other's transmission ranges. This is sometimes known as a *primary collision* and implies that $u$ and $v$ should not broadcast using the same frequency. Figure 5.17(b), meanwhile, denotes a *secondary* collision . Here, although there is no primary collision between devices $v_1$ and $v_2$, it is still necessary that they broadcast at different frequencies in order to allow $u$ to be able to distinguish between them.

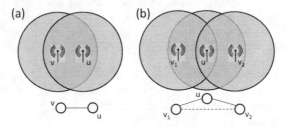

**Fig. 5.17** Illustration of a primary collision (a), and (b) a secondary collision (dotted line) in a wireless network

The problem of choosing suitable frequencies for each device in a wireless network can be modelled as a graph colouring problem by relating each device to a vertex, with edges then occurring between any pair of vertices subject to a primary and/or secondary collision. Each frequency then corresponds to a colour and, due to the finite number of frequencies that exist in the radio spectrum, we now wish to colour this graph using the minimal number of colours.

More precisely, let $G = (V,E)$ be a graph with vertex set $V$ and an edge $E$. The set of edges due to primary collisions, $E_1$, contains all pairs of devices that are close enough to be able receive each other's transmissions (as with Figure 5.17(a)). The set $E_2$ then contains pairs of devices subject to secondary collisions: that is, $\{v_i, v_j\} \in E_2$ if and only if the distance between $v_i$ and $v_j$ in the graph $G_1 = (V,E_1)$ is exactly 2 (as is the case in Figure 5.17(b)). If only primary collisions need to be considered when assigning frequencies, we only need to colour the graph $G_1$; otherwise we will need to colour the graph $G = (V, E = E_1 \cup E_2)$. In either case, this task is a type of decentralised graph colouring problem because each vertex (wireless device) is responsible for choosing its own colour (frequency), while being aware only of its neighbours and their current colours.

One simple but effective algorithm for the decentralised graph colouring problem is due to Finocchi et al. (2005). Let $G = (V,E)$ be a graph with maximal degree $\Delta(G)$. To begin, all vertices in $G$ are set as uncoloured. Each vertex is also allocated a set of candidate colours, defined $L_v = \{1, 2, \ldots, \deg(v) + 1\}\ \forall v \in V$. A single iteration of the algorithm then involves the following four steps.

1. In parallel, each uncoloured vertex $v$ selects a *tentative* colour $t_v \in L_v$ at random.
2. In parallel, consider each tentatively coloured vertex $v$. If no neighbour of $v$ is coloured with $t_v$, then set $t_v$ as the *final* colour of $v$.
3. In parallel, consider each remaining tentatively coloured vertex $v$ and

   a. Remove its tentative colour.
   b. Update $L_v$ by deleting all colours from $L_v$ that are assigned as final colours to neighbours of $v$.
   c. If $L_v = \emptyset$ then let $l$ be the largest colour label assigned as a final colour in $v$'s neighbourhood and set $L_v = \{1, 2, \ldots, \min\{l + 1, \Delta(G) + 1\}\}$.

4. If any uncoloured vertices remain, return to Step 1.

An example run of this algorithm is shown in Figure 5.18. In the first iteration, we see that three of the five vertices are allocated final colours; the remaining two vertices are then allocated final colours in the second iteration.

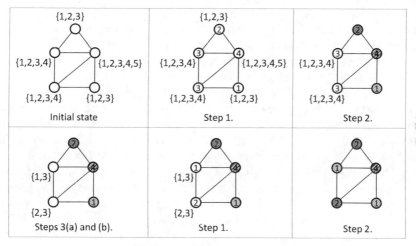

Fig. 5.18 Example run of Finocchi et al.'s algorithm. Here tentatively coloured vertices appear in white, with labels within the vertices indicating tentative colours

Note that in the above algorithm, each vertex $v$ is initially assigned a set of candidate colours $L_v = \{1, 2, \ldots, \deg(v) + 1\}$. This means that $L_v$ always contains sufficient options to allow each vertex $v$ to be coloured differently from all of its neighbours; hence Step 3(c) will never actually be used. If, however, we desire a solution using fewer colours, we might choose to introduce a *shrinking factor* $s > 1$, which can be used to limit the initial set of candidate colours for each vertex $v$ to

$L_v = \{1, 2, \ldots, \lfloor \frac{\deg(v)+1}{s} \rfloor\}$. In this case Step 3(c) might now be needed if the original contents of $L_v$ prove insufficient. The algorithm may also need to execute for an increased number of iterations in order to achieve a feasible solution.

Finocchi et al. (2005) also suggest an improvement to this algorithm by replacing Step 2 with a more powerful operator. Observe in Step 2 of the first iteration of Figure 5.18 that there are two vertices tentatively coloured with colour 3. Accordingly, neither of these vertices receives a final colour at this iteration, though it is obvious that one of them could indeed receive colour 3 as a final colour at this point. An improvement to Step 2 therefore operates as follows. Let $G(i) = (V(i), E(i))$ be the subgraph induced by all vertices tentatively coloured with colour $i$. We now identify a maximal independent set for $G(i)$ and assign all vertices in this set to a final colour $i$. All other vertices in $G(i)$ remain uncoloured. To form this independent set, in parallel each vertex $v \in V(i)$ first generates a random number $r_v \in [0, 1]$. The tentative colour of a vertex $v$ is then selected as its final colour if and only if $r_v$ is less than the random numbers chosen by its neighbours in $G(i)$. This is equivalent to the greedy process of randomly permuting the vertices in $V(i)$ and then adding each vertex $v \in V(i)$ to the independent set if and only if it appears before its neighbours in the permutation.

Decentralised graph colouring problems arise in a number of practical situations, including TDMA slot assignment, wake-up scheduling and data collection (Hernández and Blum, 2014). One noteworthy piece of research is due to Kearns et al. (2006), who have examined the decentralised colouring of graphs representing social networks. In their case each vertex in the graph is a human participant, and two vertices are adjacent only if these people are judged to know one another. The objective of the problem is for each person to choose a colour for himself or herself only by using information regarding the colours of his or her neighbours. Participants are also able to change their colour as often as necessary until, ultimately, a feasible colouring of the entire graph is formed. This problem has real-world implications in situations where it is desirable to distinguish oneself from one's neighbours: for example, selecting a mobile phone ringtone that differs from one's friends, or choosing to develop professional expertise that differs from one's colleagues. In their research, Kearns et al. (2006) carried out a number of experiments on various graph topologies using segregated participants. Under a time limit of five minutes, topologies such as cycle graphs were optimally coloured quite quickly through the collective efforts of the participants. Other, more complex graphs modelling more realistic social network topologies were seen to present more difficulties, however.

## 5.6.2 Online Graph Colouring

In the online graph colouring problem, a graph is gradually revealed to an algorithm by presenting the vertices one at a time. The algorithm must then assign each vertex to a colour before receiving the next vertex in the sequence. In other words, an online graph colouring algorithm receives vertices in a given ordering $v_1, v_2, \ldots, v_n$ and the colour for a vertex $v_i$ is determined by only considering the colours of the vertices in the subgraph induced by the set $\{v_1, \ldots, v_i\}$. Once $v_i$ has been coloured, it generally cannot be changed by the algorithm.

Much of the research relating to online graph colouring looks at the worst case behaviour of algorithms on particular topologies. Let $A$ be an online graph colouring algorithm and consider the colourings of a graph $G$ produced by $A$ over all orderings of $G$'s vertices. The maximum number of colours used among these colourings is denoted by $\chi_A(G)$. That is, $\chi_A(G)$ denotes the worst possible performance of $A$ on $G$.

It has been shown by Lovász et al. (1989) that if $G$ is a bipartite graph with $n$ vertices, then there exists an online algorithm $A$ such that

$$\chi_A(G) \leq 2\log_2 n. \tag{5.2}$$

Kierstead and Trotter (1981) have also shown that if $G$ is an interval graph then there exists an online algorithm $A$ such that

$$\chi_A(G) \leq 3\chi(G) - 2. \tag{5.3}$$

Studies of online colouring have also focussed on the behaviour of the GREEDY algorithm, which, we recall, operates by assigning each vertex to the lowest indexed colour seen to be feasible (see Section 2.1). Bounds noted by Gyárfás and Lehel (1988) include

$$\chi_{\text{GREEDY}}(G) \leq \chi(G) + 1 \tag{5.4}$$

if $G$ is a split graph (i.e., a graph that can be partitioned into one clique and one independent set),

$$\chi_{\text{GREEDY}}(G) \leq \frac{3}{2}\chi(G) + 1 \tag{5.5}$$

if $G$ is the complement of a bipartite graph, and

$$\chi_{\text{GREEDY}}(G) \leq 2\chi(G) - 1 \tag{5.6}$$

if $G$ is the complement of a chordal graph.

Empirical work by Ouerfelli and Bouziri (2011) has also suggested that instead of following the GREEDY algorithm's strategy of assigning vertices to the lowest indexed feasible colour, it is often beneficial to assign vertices to the feasible colour containing the *most* vertices. This is because such a heuristic will often aid the formation of larger independent sets in a solution, ultimately helping to reduce the number of colours used in the final solution.

A real-world application of online graph colouring is presented by Dupont et al. (2009). Here, a military-based frequency assignment problem is considered in which wireless communication devices are introduced one by one into a battlefield environment. From a graph colouring perspective, given a graph $G = (V, E)$, the problem starts with an initial colouring of the subgraph induced by the subset of vertices $\{v_1, \ldots, v_i\}$. The remaining vertices $v_{i+1}, \ldots, v_n$ are then introduced one at a time, with the colour (frequency) of each vertex having to be determined before the next vertex in the sequence is considered. In this application the number of available colours is fixed from the outset, so it is possible that a vertex $v_j$ $(i < j \leq n)$ might be introduced for which no feasible colour is available. In this case a repair operator is invoked that attempts to rearrange the existing colouring so that a feasible colour is created for $v_j$. Because such rearrangements are considered expensive, the repair operator also attempts to minimise the number of vertices that have their colours changed during this process.

### 5.6.3 Dynamic Graph Colouring

Dynamic graph colouring differs from decentralised and online graph colouring in that we again possess a global knowledge of the graph we are trying to colour. However, in this case a graph is also able to randomly change over time via the deletion and introduction of vertices, and/or the deletion and introduction of edges. We must therefore re-solve the problem at each stage.

A practical application of dynamic graph colouring might occur in the timetabling of lectures at a university (see Section 1.1.2 and Chapter 8). To begin, a general set of requirements and constraints will be specified by the university and an initial timetable will be produced. However, on viewing this draft timetable, circumstances might dictate that some constraints need to be altered, additional lectures need to be introduced, or other lectures need to be cancelled. This will result in a new timetabling problem that needs to be solved, with the process continuing in this fashion until a finalised solution is agreed upon.

More generally, a dynamic graph colouring problem can be defined by a sequence of graphs $\mathcal{G} = (G_1, G_2, \ldots, G_{|\mathcal{G}|})$, where a solution to each graph $G_i \in \mathcal{G}$ will need to be produced within a limited time frame before the next graph $G_{i+1}$ is considered. One of the main issues to consider here is the level of similarity between two successive graphs $G_i$ and $G_{i+1}$ in this sequence. If $G_i$ and $G_{i+1}$ are seen to be quite alike, then it may be beneficial to use the solution generated for $G_i$ as a starting solution for $G_{i+1}$. However, if the differences are larger, then it may be more appropriate to simply colour $G_{i+1}$ from scratch.

## 5.7 List Colouring

The list colouring problem, like the (vertex) graph colouring problem, involves assigning colours to each vertex of a graph such that no pair of adjacent vertices are assigned to the same colour. However, in addition to this, when a problem is specified, individual vertices are also allocated their own list of permissible colours to which they can be assigned.

Defined more precisely, the list colouring problem gives us a graph $G = (V, E)$ and also a set $L_v$ of permissible colours for each vertex $v \in V$. The sets $L_v$ are usually referred to as a "lists", giving the problem its name. The task is to now produce a feasible colouring of $G$ (that is, a colouring that is both proper and complete), with the added restriction that all vertices only be coloured using colours appearing in their corresponding lists: that is, $\forall v \in V, c(v) \in L_v$. If a $k$-colouring exists for a particular problem instance of the list colouring problem, we say that the graph $G$ is "$k$-chooseable". The "choice number" $\chi_L(G)$ then refers to the minimum $k$ for which $G$ is $k$-chooseable. Note that the chromatic number of a graph $\chi(G) \leq \chi_L(G)$.

List colouring problems have obvious applications in areas such as timetabling where, in addition to scheduling events into a minimal number of timeslots (as we saw in Section 1.1.2), we might also face constraints of the form "event $v$ can only be assigned to timeslots $x$ and $y$", or "event $u$ cannot be assigned to timeslot $z$". List colouring, stated as a decision problem, is also NP-complete because it generalises the graph colouring problem itself. That is, all $k$-colouring problems can be easily converted into an equivalent list colouring decision problem of deciding whether $G$ is $k$-chooseable by simply setting $L_v = \{1, 2, \ldots, k\}, \forall v \in V$.

In practice, algorithms for the graph colouring problem can often be used for deciding whether a graph is $k$-chooseable, beyond those for which $L_v = \{1, 2, \ldots, k\}, \forall v \in V$. More specifically, graph colouring algorithms can be used to tackle any list colouring problem for which our chosen $k \geq |L|$, where $L$ is defined as the union of all lists: $L = \bigcup_{v \in V} L_v$. Imagine we have a list colouring problem defined on a graph $G$ for which $k \geq |L|$ is satisfied. First, we create a new graph $G'$ by copying the vertices and edges of $G$ and then adding $k$ additional vertices, which we label $u_1, u_2, \ldots, u_k$. Next we add edges between all $\binom{k}{2}$ pairs of these additional vertices so that they form a complete graph $K_k$. This implies that any feasible colouring of $G'$ must use at least $k$ different colours. Without loss of generality, we can assume that $c(u_i) = i$ for $1 \leq i \leq k$. Finally we then go through each vertex $v$ in $G'$ that came from the original graph $G$ and consider its colour list, adding an edge between $v$ and $u_i$ if colour $i \notin L_v$. This has the effect of disallowing $v$ from being assigned to colour $i$ as required.

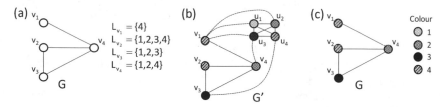

**Fig. 5.19** Illustration of how a list colouring problem (a) can be converted into an equivalent graph colouring problem (b), whose colouring then represents a feasible solution to the original list colouring problem (c)

Figure 5.19 demonstrates this process. In this example $k = |L| = 4$, meaning that four additional vertices $u_1, \ldots u_4$ are added (larger values for $k$ are also permitted). The colouring produced for the extended graph $G'$ also uses four colours, which is the chromatic number in this case. However, we also observe that none of the vertices originating from $G$ are actually coloured with colour 1 in this example; hence we deduce that $G$ is actually 3-chooseable, as shown in Figure 5.19(c).

## 5.8 Equitable Graph Colouring

Another extension to the graph colouring problem is the *equitable* colouring problem, where we seek to establish a feasible colouring of a graph $G$ such that the sizes of the colour classes differ by at most 1. In other words, we seek to establish a feasible $k$-colouring so that exactly $n \bmod k$ colour classes contain $\lceil n/k \rceil$ vertices, and the remainder contain exactly $\lfloor n/k \rfloor$ vertices.

Examples of equitable graph colouring problems occur quite naturally as extensions to the general graph colouring problem. In university timetabling for example, it might be desirable to minimise the number of rooms required by balancing the number of events per timeslot (see Section 1.1.2). Another application can be found in the creation of table plans for large parties. Imagine, for example, that we have $n$ guests who are to be seated at $k$ equal-sized tables, but that some guests are known to dislike each other and therefore need to be assigned to different tables. In this case we can model the problem as a graph by using vertices for guests, with edges occurring between pairs of guests who dislike each other. An extension to this application is the subject of Chapter 6.

Let $G = (V, E)$ be a graph with $n$ vertices, a maximal degree $\Delta(G)$, and an independence number $\alpha(G)$.

**Definition 5.6** *If $V$ can be partitioned into $k$ colour classes $S = \{S_1, \ldots, S_k\}$ such that $S_i$ is an independent set and $||S_i| - |S_j|| \leq 1 \ \forall i \neq j$, then $S$ is said to be an equitable $k$-colouring of $G$.*

**Definition 5.7** *The smallest $k$ for which an equitable $k$-colouring of $G$ exists is the equitable chromatic number, denoted by $\chi_e(G)$.*

Like the graph colouring problem, the equitable graph colouring problem is known to be NP-complete. This follows from the fact that the problem of deciding whether a graph $G$ is $k$-colourable can be converted into an equitable $k$-colouring problem by simply adding an appropriate number of isolated vertices to $G$. Hence the equitable graph colouring problem generalises the standard graph colouring problem. An alternative proof is also due to Furmańczyk (2004), who shows that the problem of deciding whether $\chi_e(G) \leq 3$ is NP-complete, even when $G$ is the line graph of a cubic graph.

Because a feasible equitable $k$-colouring of a graph $G$ is also a feasible $k$-colouring of $G$, it is obvious that $\chi(G) \leq \chi_e(G)$. In some cases however, this bound can be very poor. Consider star graphs, for example, which comprise a vertex set $V = \{v_1, \ldots, v_n\}$ and an edge set $E = \{\{v_1, v_i\} : i \in \{2, \ldots, n\}\}$. These are a type of bipartite graph and therefore feature a chromatic number of 2. However, it is obvious that $v_1$ must be assigned a different colour to all other vertices; hence, in an equitable colouring all other colour classes must contain a maximum of two vertices. The equitable chromatic numbers of star graphs are therefore calculated as $\lceil (n-1)/2 \rceil + 1$, as illustrated in Figure 5.20.

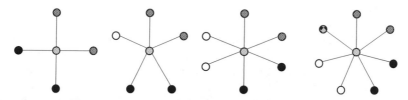

**Fig. 5.20** The equitable chromatic numbers for star graphs with $n = 5, 6, 7$, and 8 are 3, 4, 4, and 5 respectively

A better lower bound for equitable chromatic numbers on general graphs is outlined by Furmańczyk (2004). Suppose that $G$ is equitably coloured, with vertex $v$ assigned to colour 1. The number of vertices coloured with colour 1 is therefore at most $\alpha(G - \Gamma(v) - \{v\}) + 1$. Since this colouring is equitable, the number of vertices coloured with any other colour is then at most $\alpha(G - \Gamma(v) - \{v\}) + 2$. Hence:

$$\left\lceil \frac{n+1}{\alpha(G - \Gamma(v) - \{v\}) + 2} \right\rceil \leq \chi_e(G) \tag{5.7}$$

For example, using a star graph with $n = 8$ whose internal vertex $v_1$ is coloured with colour 1, this gives $\left\lceil \frac{8+1}{0+2} \right\rceil = 5$. Note, however, that this bound requires a graph's independence number to be calculated, which is itself an NP-hard problem.

With regard to upper bounds on $\chi_e(G)$, it is known that any graph can be equitably $k$-coloured when $k \geq \Delta(G) + 1$. Hence:

**Theorem 5.15 (Hajnal and Szemerédi (1970))** *Let $G$ be a graph with maximal degree $\Delta(G)$. Then $\chi_e(G) \leq \Delta(G) + 1$.*

This fact was initially conjectured by Erdős (1964), with a formal proof being published six years later by Hajnal and Szemerédi (1970). Shorter proofs of this theorem have also been shown by Kierstead and Kostochka (2008) and Kierstead et al. (2010). The latter publication also presents a polynomial-time algorithm for constructing an equitable $(\Delta(G) + 1)$-colouring. This method involves first removing all edges from $G$ and dividing the $n$ vertices arbitrarily into $\Delta(G)$ equal-sized colour classes. In cases where $n$ is not a multiple of $\Delta(G)$, sufficient isolated vertices are added. The vertices are then considered in turn and, in each iteration $i$, the edges incident to vertex $v_i$ are added to $G$. If $v_i$ is seen to be adjacent to another vertex in its colour class, it is moved to a different feasible colour class, leading to a feasible colouring with up to $\Delta(G) + 1$ colours. If this colouring is not equitable, then a

polynomial-length sequence of adjustments are made to reestablish the balance of
the colour classes.

It is notable that Theorem 5.15 above is similar to Theorem 2.5 from Chapter 2
which states that for any graph $G$, $\chi(G) \leq \Delta(G) + 1$. Meyer (1973) has gone one
step further to even conjecture a form of Brooks' Theorem (2.7) for equitable graph
colouring: every graph $G$ has an equitable colouring using $\Delta(G)$ or fewer colours
with the exception of complete graphs and odd cycles. Recall, however, that the
problem of determining an equitable $k$-colouring for an arbitrary $k$ and graph $G$ is
still NP-complete, implying the need for approximation algorithms and heuristics
in general. One simple approach for achieving approximate equitable $k$-colourings
(for $k \leq \Delta(G)$) can be achieved via a simple modification of the DSATUR algorithm:
starting with $k$ empty colour classes, take each vertex in turn according to DSATUR's
heuristics and assign it to the feasible colour class containing the fewest vertices,
breaking ties randomly.

Figure 5.21 summarises results achieved by this algorithm for random graphs
$G_{500,p}$, using $p = 0.1, 0.5$ and $0.9$, for a range of suitable $k$-values. For comparison's
sake, the results of a second algorithm are also included here. This operates in the
same manner except that vertices are assigned to a randomly chosen feasible colour.
The cost here is simply the difference in size between the largest and smaller colour
classes in a solution. Hence a cost of 0 or 1 indicates an equitable $k$-colouring.

Figure 5.21 clearly demonstrates that, for these graphs, the policy of assigning
vertices to feasible colour classes with the fewest vertices brings about more equitably coloured solutions. We also see that the algorithm consistently achieves equitable colourings for the majority of $k$-values with the exception of those close to the
chromatic number, and those where $k$ is a divisor of $n$. For the former case, the low
number of available colours restricts the choice of feasible colours for each vertex,
often leading to inequitable colourings. On the other hand, when $k$ is a divisor of
$n$ the algorithm is seeking a solution with a cost of 0, meaning that the last vertex
considered by the algorithm must be assigned to the unique colour class containing one fewer vertex than the remaining colour classes. If this colour turns out to
be infeasible (which often seems to be the case), this vertex will then need to be
assigned to another colour class, resulting in a solution with a cost of 2. Note, however, that it might be possible to further improve these solutions by, for example,
applying a local search algorithm with appropriate neighbourhood operators such
as Kempe chain interchanges and pair swaps. An approach along these lines for a
related problem is the subject of the case study presented in Chapter 6.

## 5.9 Weighted Graph Colouring

Further useful extensions of the graph colouring problem can be achieved through
the addition of numeric *weights* to a graph. Typically, the term "weighted graph
colouring" is used in situations where the vertices of a graph are allocated weights.
However, the term is also sometimes used for problems where edges are weighted,

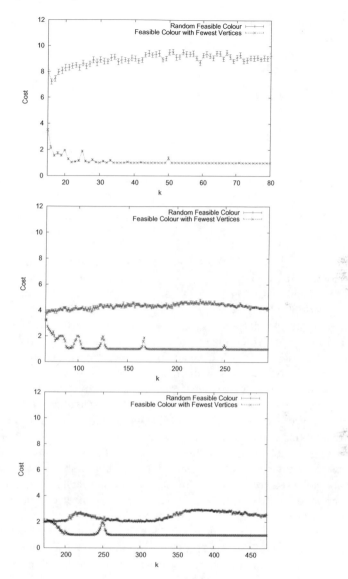

**Fig. 5.21** Quality of equitable solutions produced by the modified DSATUR algorithms on random graphs with $n = 500$ for, respectively, $p = 0.1$, $0.5$ and $0.9$. All figures, are the average of 50 instances per $k$-value. Error bars show one standard error either side of these means

and for the multicolouring problem. These are considered in turn in the following subsections.

### 5.9.1 Weighted Vertices

A natural formulation of the weighted graph colouring problem is as follows. Let $G(V,E,w)$ be a graph for which each vertex $v \in V$ is given a nonnegative integer weight $w_v$. Given a fixed number of colours $k$, our task is to identify a proper (but perhaps partial) solution $S = \{S_1, \ldots, S_k\}$ that maximises the objective function:

$$f(S) = \sum_{i=1}^{k} g(S_i) \tag{5.8}$$

where $g(S_i) = \sum_{v \in S_i} w_v$. For this problem, if $\chi(G) \leq k$ then an optimal solution will obviously feature a cost of $\sum_{v \in V} w_v$, corresponding to a feasible graph colouring solution. On the other hand, if $k < \chi(G)$, then the problem involves finding an optimal subset of vertices $V' \subset V$, where $V' = \bigcup_{i=1}^{k} S_i$.

It is straightforward to show that this formulation is NP-hard by noting that any instance of the graph colouring problem can be transformed into this weighted variant by setting $w_v = 1 \; \forall v \in V$. If, in addition to this, $k = 1$, then the problem is also equivalent to the NP-hard maximum independent set problem.

This formulation of the weighted graph colouring problem arises in practical situations where a limited number of colour classes are available and where the colouring of certain vertices is considered more important than others. For example, in an exam timetabling problem we may be given a fixed number of timeslots (colours), and we might want to prioritise the assignment of larger exams (higher weighted vertices) to the timetable while also making sure that clashing exams (adjacent vertices) are not assigned to the same timeslots. One simple heuristic for this problem is to employ the GREEDY algorithm using a fixed number of colours $k$ and an ordering of the vertices $v_1, v_2, \ldots, v_n$ such that $w_{v_1} \geq w_{v_2} \geq \ldots \geq w_{v_n}$. Algorithms that explore the space of partial proper solutions are also naturally suited. For example, we might make use of the PARTIALCOL algorithm while seeking to minimise the objective function $\sum_{v \in U} w_v$, where $U = V - \bigcup_{i=1}^{k} S_i$ is the set of uncoloured vertices (see Section 4.1.2).

Another well-known formulation of the weighted graph colouring problem involves taking a graph $G(V,E,w)$ as above, and then determining a feasible colouring $S = \{S_1, \ldots, S_k\}$ that minimises Equation (5.8) using $g(S_i) = \max\{w_v : v \in S_i\}$. A practical example of this occurs in the scheduling of fixed-time jobs to timeslots. Imagine, for example, that we are given a set of jobs $V$, each with a processing time $w_v \; \forall v \in V$. Imagine further that these jobs are to be scheduled into $k$ timeslots, and that the jobs assigned to a particular timeslot will be carried out simultaneously; hence, the *duration* of a timeslot $S_i$ is set at $\max\{w_v : v \in S_i\}$. Finally, also suppose that some pairs of jobs $u, v$ are *incompatible* and cannot be assigned to the same timeslot. Such pairs correspond to edges $\{u, v\} \in E$.

In this formulation it is obviously in our interest to try and assign vertices with large weights to the same colour classes. Similarly, it also makes sense in many cases to reduce the number of colours being used by increasing the number of colour classes $S_i$ for which $S_i = \emptyset$. Demange et al. (2007), however, have noted that the optimal number of nonempty colour classes for a particular graph $G$ may well be larger than $\chi(G)$, though it will also always be less than or equal to $\Delta(G) + 1$.

As with the previous example, this problem formulation is seen to be NP-hard by observing that any instance for which $w_v = 1$ ($\forall v \in V$) is equivalent to the standard graph colouring problem. Furthermore, the problem remains NP-hard even for interval graphs (Escoffier et al., 2006) and bipartite graphs (Demange et al., 2007). For the bipartite case, Demange et al. (2007) have provided an algorithm with approximation ratio of $4r_w/(3r_w + 2)$ (where $r_w = \frac{\max\{w_v : v \in V\}}{\min\{w_v : v \in V\}}$). They also prove that optimal solutions for bipartite graphs can be found in polynomial time whenever $|\{w_v : v \in V\}| \leq 2$. For general graphs, an approximation algorithm is also suggested that operates as follows. As before, let $G = (V, E, w)$ be a graph with weighted vertices and $g(S_i) = \max\{w_v : v \in S_i\}$.

1. Construct a graph with weighted edges $\bar{G} = (V, E', w')$ where $\bar{G}$ is the complement of $G$, and for any edge $\{u, v\} \in E'$, $w'_{uv} = w_u + w_v - g(\{w_u, w_v\})$.
2. Compute a maximum weighted matching $M^*$ of $\bar{G}$.
3. For each edge in $M^*$, colour the end points with a new colour.
4. Colour any remaining vertices with their own new colour.

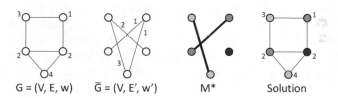

$G = (V, E, w)$ $\quad$ $\bar{G} = (V, E', w')$ $\quad$ $M^*$ $\quad$ Solution

**Fig. 5.22** Illustration of the algorithm of Demange et al. (2007)

An example of this process is shown in Figure 5.22. The matching $M^*$ can be determined in polynomial time using methods such as the blossom algorithm (Kolmogorov, 2009). Note that each colour class in the solution is an independent set, but that these are limited to contain a maximum of two vertices. Indeed, in graphs where no independent set contains more than two vertices (such as the complement of a bipartite graph), this algorithm guarantees the optimal. In further work, Hassin and Monnot (2005) have shown that, for any graph, this process produces a solution whose objective function value never exceeds twice the optimum. They also show that the same approximation ratio applies when $g(S_i)$ takes other forms such as $g(S_i) = \min\{w_v : v \in S_i\}$ and $g(S_i) = \frac{1}{|S_i|}\sum_{v \in S_i} w_v$. Malaguti et al. (2009) have also proposed a number of IP-based methods for this problem similar in spirit to those discussed in Section 3.1.2. In particular, they propose the use of heuristics for building up a large sample of independent sets, and then use an IP model similar to that of Section 3.1.2.1 to select a subset of these. Local search-based methods based on Kempe chain interchanges and pair swaps also seem to be naturally suited to this problem.

## 5.9.2  Weighted Edges

In many cases it is more convenient to apply weights to the *edges* of a graph as opposed to the vertices, allowing us to express levels of preference for assigning vertices to different (or the same) colours. One interpretation involves taking a graph $G(V, E, w)$ for which each edge $\{u, v\} \in E$ is allocated an integer weight $w_{uv}$. Given a fixed number of colours $k$, our task is to then identify a complete (but perhaps improper) solution $S = \{S_1, \ldots, S_k\}$ that minimises the objective function:

$$f(S) = \sum_{i=1}^{k} \sum_{u,v \in S_i : \{u,v\} \in E} w_{uv} \tag{5.9}$$

Here, if $\forall \{u, v\} \in E$, $w_{uv} > 0$, then any solution for which $f(S) = 0$ corresponds to a feasible $k$-coloured solution.

This sort of formulation is applicable in areas such as university timetabling and social networking. For the former, imagine, as before, that we wish to assign events (vertices) to timeslots (colours), but that there are insufficient timeslots to feasibly accommodate all events. In order to form a complete timetable, this means that some clashes will be necessary; however, some types of clashes may be deemed less critical than others. For example, if two clashing events only have a small number of common students, then we may allow them to both be assigned to the same timeslots (with alternative arrangements then being made for the people affected). On the other hand, if two events contain a large number of common participants, or if the same instructor is required to teach them both, then such a clash would be far less desirable. Appropriate weights added to the corresponding edges can be used to express such preferences.

Note that due to the nature of this problem's requirements, algorithms that search the space of complete improper solutions will often be naturally suitable here. In Chapter 6 an application along these lines will be made to the problem of partitioning members of social networks, where edge weights are used to express a level of "liking" or "disliking" between pairs of individuals.

### 5.9.3 *Multicolouring*

Another problem that is sometimes referred to as "weighted graph colouring" is the NP-hard graph *multicolouring* problem. In this case we are given a graph $G(V, E, w)$ for which each vertex $v \in V$ is allocated a weight $w_v \in \{1, 2, \ldots\}$. The task is to then assign $w_v$ different colours to each vertex $v$ such that (a) adjacent vertices have no colours in common, and (b) the number of colours used is minimal.

Multicolouring has practical applications in areas such as frequency assignment problems where, in some cases, it is desirable for devices to be able to transmit and receive messages on multiple frequencies as opposed to just one (Aardel et al., 2002). McDiarmid and Reed (2000) have shown that this problem is polynomially solvable for bipartite and perfect graphs, but that it remains NP-hard for triangular lattice graphs and their induced subgraphs, which have practical applications in cellular telephone networks. They also suggest a suitable polynomial-time approximation algorithm for the latter topology.

Note that the graph colouring problem is a special case of the multicolouring problem for which $w_v = 1 \; \forall v \in V$. On the other hand, any instance of the multicolouring problem can also be converted into an equivalent graph colouring problem by replacing each vertex $v \in V$ with a clique of size $w_v$, and then connecting every member of the clique to all neighbours of $v$ in $G$. This method of conversion allows us to use any graph colouring algorithm (such as those from Chapter 4) with the graph multicolouring problem, though it also increases the number of vertices to colour by a factor of $\sum_{v \in V} w_v / n$. Consequently, graph multicolouring is often studied as a separate computational problem, for which the backtracking algorithm of Caramia and Dell'Olmo (2001) and the IP branch-and-price method of Mehrotra and Trick (2007) are prominent examples.

# Chapter 6
# Designing Seating Plans

Chapters 6, 7 and 8 each contain a detailed case study showing how graph colouring methods can be used to successfully tackle important real-world problems. The first of these case studies concerns the task of designing table plans for large parties, which, as we will see, combines elements of the NP-hard (edge) weighted graph colouring problem, the equitable graph colouring problem and the $k$-partition problem. A user-friendly implementation of the algorithm proposed in this section can also be found online at www.weddingseatplanner.com.

## 6.1 Problem Background

Consider a social event such as a wedding where, as part of the formalities of the day, $N$ guests need to be divided among $k$ dining tables. To ensure that guests have a good time it will often be necessary to design a seating plan, thereby allowing guests to be seated at tables with appropriate company. The following sorts of factors might be taken into account:

- Guests belonging to groups, such as couples and families with small children, should be seated at the same tables, preferably next to each other.
- If there is any perceived animosity between different guests, they should be seated at different tables. Similarly, if guests are known to enjoy one anothers' company, it may be desirable for them to be seated at the same table.
- Some guests might be required to sit at a particular table (e.g. close to the kitchen or washrooms). Similarly, some guests might be prohibited from sitting at certain tables.
- Since tables could vary in size and shape, each table should be allocated a suitable number of guests, and these guests should be arranged around the table in an appropriate manner.

A naïve method for producing a seating plan best fitting these sorts of criteria might be to consider all possible plans and then choose the one perceived to be the most

© Springer International Publishing Switzerland 2016
R.M.R. Lewis, *A Guide to Graph Colouring*,
DOI 10.1007/978-3-319-25730-3_6

suitable. However, for nontrivial values of $N$ or $k$, the number of possible solutions is too large for this to be possible. To illustrate, consider a simple example where we have $N = 48$ guests using $k = 6$ equal sized tables (i.e., exactly eight guests per table). For the first table we need to choose eight people from the 48, for which there are $\binom{48}{8} = 377,348,994$ possible choices. For the next table, we then need to choose eight further guests from the remaining 40, giving $\binom{40}{8} = 76,904,685$ further choices, and so on. Assuming that $k$ is a divisor of $N$, using equal sized tables, the number of possible plans is thus:

$$\prod_{i=0}^{k-2} \binom{N - i(N/k)}{N/k}. \tag{6.1}$$

This function clearly has a growth rate that is subject to a combinatorial explosion—even for the modestly sized example above, the number of distinct solutions to check is

$$\binom{48}{8} \cdot \binom{40}{8} \cdot \binom{32}{8} \cdot \binom{24}{8} \cdot \binom{16}{8} \approx 3.8 \times 10^{33}$$

which is far beyond the capabilities of any state-of-the-art computing equipment. Furthermore, if we were to relax the problem by allowing $k$ tables of any size $\geq 1$, the task would now be to partition the $N$ guests into $k$ nonempty subsets, meaning that the number of solutions would be equal to a Stirling number of the second kind (Equation (1.6)). These numbers feature even higher growth rates than Equation (6.1), giving even larger solution spaces.

Such features demonstrate that this sort of naïve method for producing a desirable seating plan is clearly infeasible in most cases. However, the problem of constructing seating plans is certainly important since (a) it is regularly encountered by party organisers and (b) the quality of the proposed solution could have a significant effect on the success (or failure) of the gathering.

Currently there is a small amount of commercial software available for constructing seating plans, such as Perfect Table Plan, Top Table Planner, and Seating Arrangement.[1] The first of these examples allows users to input a list of guest names into the system and then specify preferences between these guests (such as whether they need to be seated apart or together). It then allows users to define table shapes, sizes and locations, before assisting the user in placing the guests at these tables via drag and drop functionality and also an auto assign tool. The exact details of the underlying algorithm used with the auto assign tool are not made public by the software vendor, though its online documentation states that an evolutionary algorithm is used, with different penalty costs being applied for different types of constraint violation. The fitness function of the algorithm is simply an aggregate of these penalties.

---

[1] Refer to the websites www.perfecttableplanner.com, www.toptableplanner.com, and www.seatingarrangement.com respectively.

### 6.1.1 Relation to Graph Problems

Given $N$ guests, in its simplest form the problem of constructing a seating plan can be described using a binary matrix $\mathbf{W}_{N \times N}$, where element $W_{ij} = 1$ if guests $i$ and $j$ are required to be seated apart and $W_{ij} = 0$ otherwise. We can also assume that $W_{ij} = W_{ji}$. Given this input matrix, our task might then be to partition the $N$ guests into $k$ subsets $\mathcal{S} = \{S_1, \ldots, S_k\}$, such that the objective function

$$f(\mathcal{S}) = \sum_{t=1}^{k} \sum_{\forall i,j \in S_t : i<j} W_{ij} \tag{6.2}$$

is minimised.

In fact, the problem of confirming the existence of a zero-cost solution to this problem is equivalent to the decision variant of the graph colouring problem. Here, the graph $G = (V, E)$ is defined using the vertex set $V = \{v_1, \ldots, v_N\}$ and the edge set $E = \{\{v_i, v_j\} : W_{ij} = 1 \wedge v_i, v_j \in V\}$. That is, each guest corresponds to a vertex, and two vertices $v_i$ and $v_j$ are considered to be adjacent if and only if $W_{ij} = 1$. Colours correspond to tables, and we are now interested in colouring $G$ using $k$ colours.

From an alternative perspective, consider the situation where we are again given the binary matrix $\mathbf{W}$, and where we now have a subset of guests $S \in \mathcal{S}$ that have been assigned to a particular circular-shaped table. Here, we might be interested in arranging the guests onto the table such that, for all pairs of guests $i, j \in S$, if $W_{ij} = 1$ then $i$ and $j$ are not seated in adjacent seats. This problem can also be described by a graph $G = (V, E)$ for which the vertex set $V = \{v_i : i \in S\}$ and the edge set $E = \{\{v_i, v_j\} : W_{ij} = 1 \wedge v_i, v_j \in V\}$. A Hamiltonian cycle of the complement graph $\bar{G}$ defines a seating arrangement satisfying this criterion, as illustrated in the example in Figure 6.1.

**Definition 6.1** *Given a graph $G = (V, E)$, a Hamiltonian cycle is a cycle that visits each vertex exactly once.*

However, determining the existence of a Hamiltonian cycle in an arbitrary graph is also known to be an NP-complete problem (Garey and Johnson, 1979).

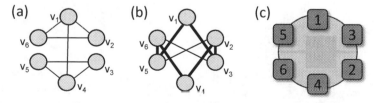

**Fig. 6.1** (a) A graph $G$ in which edges specify pairs of guests who should not be seated in adjacent seats; (b) the complement graph $\bar{G}$, together with a Hamiltonian cycle (shown in bold); and (c) the corresponding seating arrangement around a circular table

In practical situations it might be preferable for **W** to be an integer or real-valued matrix instead of binary, allowing users to place greater importance on some of their seating preferences compared to others. Assuming that smaller values for $W_{ij}$ indicate an increased preference for guests $i$ and $j$ to be seated together, the problem of partitioning the groups on to $k$ tables now becomes equivalent to the (edge) weighted graph colouring problem (Section 5.9), while the task of arranging people on to circular tables in the manner described above becomes equivalent to the travelling salesman problem. Of course, both of these problems are also NP-complete since they generalise the graph colouring problem and the Hamiltonian cycle problem respectively. In addition, the problem of arranging guests around tables could become even more complicated when tables of different shapes are used. For example, when rectangular tables are used we might also need to take account of who is seated opposite a guest in addition to his or her neighbours on either side.

### 6.1.2  Chapter Outline

In this chapter we describe a formulation of the above problem that is closely related to graph colouring, but which also involves the additional constraint of grouping together guests who like one another while also maintaining appropriate numbers of guests per table. This problem interpretation is used in conjunction with the commercial website www.weddingseatplanner.com, which contains a free tool for inputting and solving instances of the problem. The reader is invited to try out this tool while reading this chapter.

It is stated by Nielsen (2004) that users tend to leave a website in less than two minutes if it is not understood or perceived to fulfil their needs. Consequently, the particular problem formulation considered here is intended to strike the right balance between being quickly accessible to users while still being useful and flexible in practice. Since users of the website will typically have little knowledge of optimisation algorithms and the implications of problem intractability, the algorithm is also designed to supply the user with high-quality (though not necessarily optimal) solutions in very short amounts of run time (typically less than three seconds).

## 6.2  Problem Definition

In this problem formulation the $N$ guests are first partitioned into $n \leq N$ *guest groups*. Each guest group refers to a subset of guests who are required to sit together (couples, families, etc.) and will usually be known beforehand by the user. In addition to making the underlying problem smaller in most cases, specifying guest groups in this way also means that users do not have to subsequently input various preferences between pairs of people in the same families.

The Wedding Seating Problem (WSP) can now be formally stated as type of graph partitioning problem. Specifically, we are given a graph $G = (V,E)$ in which each vertex $v_i \in V$ ($i = 1,\ldots,n$) represents a guest group. The size of each guest group $v \in V$ is denoted by $s_v$, and the total number of guests in the problem is thus $N = \sum_{v \in V} s_v$.

In $G$, each edge $\{u,v\} \in E$ defines the relationship between vertices $u$ and $v$ according to a weighting $w_{uv}$ (where $w_{uv} = w_{vu}$). If $w_{uv} > 0$, this is interpreted to mean that we would prefer the guests associated with vertices $u$ and $v$ to be seated at different tables. Larger values for $w_{uv}$ reflect a strengthening of this requirement. Similarly, negative values for $w_{uv}$ mean that we would rather $u$ and $v$ were assigned to the same table.

A solution to the WSP is consequently defined as a partition of the vertices into $k$ subsets $S = \{S_1,\ldots,S_k\}$. The requested number of tables $k$ is defined by the user, with each subset $S_i$ defining the guests assigned to a particular table.

## 6.2.1 Objective Functions

Under this basic definition of the WSP, the quality of candidate solutions might be calculated according to various different metrics. In our case we use two objective functions, both that are to be minimised. The first of these is analogous to Equation (6.2),

$$f_1 = \sum_{i=1}^{k} \sum_{\forall u,v \in S_i : \{u,v\} \in E} (s_v + s_u) w_{uv}, \tag{6.3}$$

and reflects the extent to which the rules governing who sits with whom are obeyed. In this case the weighting $w_{uv}$ is also multiplied by the total size of the two guest groups involved, $s_v + s_u$. This is done so that violations involving larger numbers of people contribute more to the cost (i.e., it is assumed that $s_v$ people have expressed a seating preference in relation to guest group $u$, and $s_u$ people have expressed a preference in relation to guest group $v$).

The second objective function used in our model is intended to encourage equal numbers of guests being assigned to each table. As we noted earlier, some weddings may have varying size tables, and nearly all weddings will also have a special "head table" where the bride, groom, and their family and friends should sit. In practice, the head table and its guests can be ignored in this particular formulation because they can easily be added to the table plan once the other guests have been assigned. We also choose to assume that the remaining tables are equal-sized, which seems to be very common, particularly in larger venues. Consequently, the second objective function measures the degree to which the number of guests per table deviates from the required number of either $\lfloor N/k \rfloor$ or $\lceil N/k \rceil$ guests:

$$f_2 = \sum_{i=1}^{k} \left( \min \left( |\tau_i - \lfloor N/k \rfloor|, |\tau_i - \lceil N/k \rceil| \right) \right). \tag{6.4}$$

Here, $\tau_i = \sum_{\forall v \in S_i} s_v$ denotes the number of guests assigned to each table $i$. Obviously, if the number of guests $N$ is a multiple of $k$ then $\lfloor N/k \rfloor = \lceil N/k \rceil = N/k$, and Equation (6.4) simplifies to $f_2 = \sum_{i=1}^{k} (|\tau_i - N/k|)$.

### 6.2.2 Problem Intractability

We now show this problem to be NP-hard. We do this by showing that it generalises two classical NP-hard problems: the $k$-partition problem, and the equitable graph $k$-colouring problem. Let us first define the $k$-partition problem.

**Definition 6.2** *Let $Y$ be a multiset of $n$ weights, represented as integers, and let $k$ be a positive integer. NP-hard $k$-partition problem involves partitioning $Y$ into $k$ subsets such that the total weight of each subset is equal. (Note that the $k$-partition problem is also variously known as the load balancing problem, the equal piles problem, or the multiprocessor scheduling problem.)*

**Theorem 6.1** *The WSP is NP-hard.*

*Proof.* Let $G = (V, E)$. If $E = \emptyset$ then $f_1$ (Equation (6.3)) equals zero for all solutions. Hence, the only goal is to ensure that the number of guests per table is equal (or as close to equal as possible). Consequently the problem is equivalent to the NP-hard $k$-partition problem.

From another perspective, let $s_v = 1 \; \forall v \in V$ and let $w_{uv} = 1 \; \forall \{u, v\} \in E$. The number of guests assigned to each table $i$ therefore equals $|S_i|$. This special case is equivalent to the NP-hard optimisation version of the equitable $k$-colouring problem (see Section 5.8).                                                          □

## 6.3 Problem Interpretation and Tabu Search Algorithm

In the online tool, the user is first asked to input (or import) the names of all guests into an embedded interactive table. Guest groups that are to be seated together (families, etc.) are placed in the same rows of the table, thus defining the guest groups and values for $s_v$. Guests to be seated at the top table are also specified. At the next step the user is asked to define seating preferences between different guest groups. Since guests to be seated at the top table have already been input, constraints only need to be considered between the remaining guest groups.

Figure 6.2 shows a small example of this process. Here, nine guest groups ranging in size from 1 to 4 have been input, though one group of four has been allocated to the top table. Consequently, only the remaining eight groups ($N = 20$ guests) are considered. Figure 6.2(b) shows the way in which seating preferences (values for $w_{uv}$) are defined between these. On the website this is done interactively by clicking on the relevant cells in the grid. In this case, users are limited to three options: (1)

(a)

| Top table? | Guest name | Companion 1 | Companion 2 | Companion 3 |
|---|---|---|---|---|
| 1 ✓ | Cath | Michael | Kurt | Rosie |
| 2 | John | Sarah | Jack | Jill |
| 3 | Bill | June | | |
| 4 | Pat | Susan | | |
| 5 | Una | Tom | | |
| 6 | Ruth | Kevin | Gareth | |
| 7 | Ken | Frank | Bobby | |
| 8 | Rod | Dereck | Freddy | |
| 9 | Jane | | | |

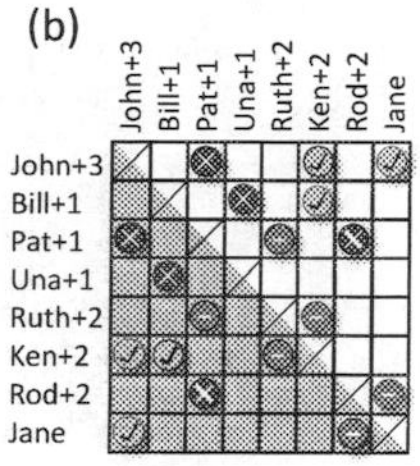


**Fig. 6.2** Specification of guest groups (left) and seating preferences (right)

"Definitely Apart" (e.g. Pat and John); (2) "Rather Apart" (Pat and Ruth); and (3) "Rather Together" (John and Ken). These are allocated weights of $\infty$, 1, and $-1$ respectively for reasons that will be made clear below. Theoretically, it is possible to allow any arbitrary weights in the grid; however, while being more flexible, it was felt by the website's interface designers that this ran the risk of bamboozling the user while not improving the effectiveness of the tool (Carroll and Lewis, 2013).

Once the input has been defined by the user, the overall strategy of our algorithm is to classify the requirements of the problem as either hard (mandatory) constraints or soft (optional) constraints. In our case we consider just one hard constraint, which we attempt to satisfy in Stage 1—specifically the constraint that all pairs of guest groups required to be "Definitely Apart" be assigned to different tables. This problem is equivalent to the graph $k$-colouring problem. In Stage 2, the algorithm then attempts to reduce the number of violations of the remaining constraints via specialised neighbourhood operators that do not allow any of the hard constraints satisfied in Stage 1 to be reviolated. The two stages of the algorithm are now described.

## 6.3.1 Stage 1

In Stage 1 the algorithm operates on the subgraph $G' = (V, E')$, where each vertex $v \subset V$ represents a guest group, and the edge set $E' = \{\{u,v\} \in E : w_{uv} = \infty\}$. In other words, the graph $G'$ contains only those edges from the original graph $G$ that define the "Definitely Apart" requirement. Using this subgraph, the problem of assigning all guests to $k$ tables (while not violating the "Definitely Apart" constraint) is equivalent to finding a feasible $k$-colouring of $G'$.

In our case, an initial solution is produced using the variant of the DSATUR heuristic used with the equitable graph colouring problem in Section 5.8. Starting with $k$ empty colour classes (tables), each vertex (guest group) is taken in turn according to the DSATUR heuristic and assigned to the feasible colour class containing the fewest vertices, breaking ties randomly. If no feasible colour exists for a vertex then it is kept to one side and is assigned to a random colour at the end of this process, thereby introducing violations of the hard constraint.

If the solution produced by the above constructive process contains hard constraint violations, an attempt is then made to eliminate them using TABUCOL (see Section 4.1.1). As we saw in Chapter 4, this algorithm can often be outperformed by other approaches in terms of the quality of solution it produces, but it does have the advantage of being very fast, which is an important requirement in this application. Consequently, TABUCOL is only run for a fixed number of iterations, specifically $20n$.

If at the end of this process a feasible $k$-colouring for $G'$ has not been achieved, $k$ is incremented by 1, and Stage 1 of the algorithm is repeated. Of course, this might occur because the user has specified a $k$-value for which no $k$-colouring exists (that is, $k < \chi(G')$) or it might simply be that a solution *does* exist, but that the algorithm has been unable to find it in the given computation limit. The process of incrementing $k$ and reapplying DSATUR and TABUCOL continues until all of the hard constraints have been satisfied, resulting in a feasible colouring of $G'$.

## 6.3.2 Stage 2

Having achieved a feasible $k$-colouring of $G' = (V, E')$ in Stage 1, Stage 2 is now concerned with eliminating violations of the soft constraints by exploring the space of feasible solutions. That is, the algorithm will make alterations to the seating plan in such a way that no violations of the "Definitely Apart" constraint are reintroduced.

Note that movements in this solution space might be restricted—indeed the space might even be disconnected—and so it is necessary to use neighbourhood operators that provide as much solution space connectivity as possible. Ideal candidates in this case are the Kempe chain interchange and the pair swap operators seen in Chapter 3 (Definitions 3.1 and 3.2). With regard to seating plans, applications of these operators have the effect of either moving one guest group from one table to another, or interchanging two subsets of guest groups between a pair of tables.

In this case, in each iteration of Stage 2, all neighbouring solutions are evaluated, and the same acceptance criteria as TABUCOL are applied. Once a move is performed, all relevant parts of the tabu list $\mathbf{T}$ are then updated to reflect the changes made to the solution. That is, all of the vertices and colours involved in the move are marked as tabu in $\mathbf{T}$. For speed's sake, in our application a fixed-size tabu tenure of 10 is used along with an iteration limit of $10n$.

### 6.3.2.1 Evaluating All Neighbours

When evaluating the cost of all neighbouring solutions in each iteration of Stage 2, considerable speedups can be achieved by avoiding situations where a particular move is evaluated more than once. Note that a Kempe chain comprising $l$ vertices can actually be generated via $l$ different vertex/colour combinations.

For example, the Kempe chain $\{v_4, v_7, v_8, v_9\}$ depicted in Figure 6.3 corresponds to KEMPE$(v_4, 1, 2)$, KEMPE$(v_7, 2, 1)$, KEMPE$(v_8, 2, 1)$, and KEMPE$(v_9, 1, 2)$. Of course, only one of these combinations needs to be considered at each iteration.

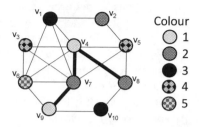

**Fig. 6.3** Example Kempe chain, KEMPE$(v_4, 1, 2) = \{v_4, v_7, v_8, v_9\}$

To achieve these speedups an additional matrix $\mathbf{K}_{n \times k}$ is used where, given a vertex $v \in S_i$, each element $K_{vj}$ is used to indicate the size of the Kempe chain formed via KEMPE$(v, i, j)$. This matrix is populated in each iteration of tabu search according to the steps shown in Figure 6.4. As can be seen here, initially all elements of $\mathbf{K}$ are set to zero. The algorithm then considers each vertex $v \in S_i$ in turn (for $1 \leq i \leq k$) and, according to line (5), only evaluates an interchange involving the chain KEMPE$(v, i, j)$ if the same set of vertices has not previously been considered. If a new Kempe chain is identified, the cost of performing this interchange is then evaluated (line (6)), and the matrix $\mathbf{K}$ is updated to make sure that this interchange is not evaluated again in this iteration (lines (7-9)).

| EVALUATE-ALL-KEMPE-CHAIN-INTERCHANGES $(\mathcal{S} - \{S_1, \ldots, S_k\})$ |
|---|
| (1) $K_{vi} \leftarrow 0 \ \forall v \in V, \ \forall i \in \{1, \ldots, k\}$ |
| (2) **forall** $S_i \in \mathcal{S}$ |
| (3)     **forall** $v \in S_i$ |
| (4)         **forall** $S_j \in (\mathcal{S} - \{S_i\})$ |
| (5)             **if** $K_{vj} = 0$ **then** |
| (6)                 Evaluate the cost of $\mathcal{S}$ if a KEMPE$(v, i, j)$ interchange were to be applied. |
| (7)                 **forall** $u \in$ KEMPE$(v, i, j)$ |
| (8)                     **if** $u \in S_i$ **then** $K_{uj} \leftarrow |\text{KEMPE}(v, i, j)|$ |
| (9)                     **else** $K_{ui} \leftarrow |\text{KEMPE}(v, i, j)|$ |

**Fig. 6.4** Procedure for efficiently evaluating all possible Kempe chain interchanges in a solution $\mathcal{S}$

Finally, after the evaluation of all possible Kempe chain interchange moves, the information in $\mathbf{K}$ can also be used to quickly identify all possible moves achievable via the pair swap operator. Specifically, for each $v \in S_i$ (for $1 \leq i \leq k - 1$) and each $u \in S_j$ (for $i + 1 \leq j \leq k$) pair swaps will only occur where both $K_{vj} = 1$ and $K_{ui} = 1$.

#### 6.3.2.2  Cost Function

The objective function used in Stage 2 of this algorithm is simply $(f_1 + f_2)$ (see Equations (6.3) and (6.4)). Note that this will always evaluate to a value less than $\infty$ since violations of the hard constraints cannot occur. Although such an aggregate function is not wholly ideal (because it involves adding together two different forms of measurement) it is acceptable in our case because in some sense both metrics relate to the number of people effected by the violations—that is, a table that is considered to have $x$ too many (or too few) people will garner the same penalty cost as $x$ violations of the "Rather Apart" constraint.

Finally, the speed of this algorithm can also be further increased by observing that (a) the cost functions $f_1$ and $f_2$ only involve the addition of terms relating to the quality of each separate colour class (table), and (b) neighbourhood moves with this algorithm only affect two colours. These features imply that if a move involving colours $i$ and $j$ is made in iteration $l$ of the algorithm, then in iteration $l+1$, the cost changes involved with moves using any pair of colours from the set $(\{1,\ldots,k\} - \{i,j\})$ will not have changed and do therefore not have to be recalculated by the algorithm.

## 6.4  Algorithm Performance

In this section we analyse the performance of our two-stage tabu search algorithm in terms of both computational effort and the costs of its resultant solutions.

The algorithm and interface described above were implemented in ActionScript 3.0, which allows the program to be executed via a web browser, with all computations being performed at the client side (note that an installation of Adobe Flash Player is required). To ensure run times are kept relatively short, and to also allow the interface to be displayed clearly on the screen, problem size is limited to $n = 50$ guest groups of up to eight people, allowing a maximum of $N = 400$ guests.

To gain an understanding of the performance characteristics of this algorithm, a set of maximum-sized problem instances of $n = 50$ guest groups (vertices) were constructed, with the size of each group chosen uniform randomly in the range 1 to 8 giving $N \approx 50 \times 4.5 = 225$. These instances were then modified such that a each pair of vertices was joined by an $\infty$-weighted edge with probability $p$, meaning that a proportion of approximately $p$ guest group pairs would be required to be "Definitely Apart". Tests were then carried out using values of $p = \{0.0, 0.3, 0.6, 0.9\}$ with numbers of tables $k = \{3, 4, \ldots, 40\}$.

Figure 6.5 shows the results of these tests with regard to the costs that were achieved by the algorithm at termination. Note that for $p \geq 0.3$, values are not reported for the lowest $k$ values because feasible $k$-colourings were not achieved (possibly because they do not exist). The figure indicates that, with no hard constraints ($p = 0.0$), balanced table sizes have been achieved for all $k$-values up to 30. From this point onwards, however, it seems there are simply too many tables (and too few

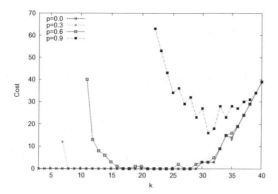

**Fig. 6.5** Solution costs for four values of $p$ using various $k$-values

guests per table) to spread the guest groups equally. Higher costs are also often incurred for larger values of $p$ because, in these cases, many guest group combinations (including many of those required for achieving low-cost solutions) will now contain at least one hard constraint violation, meaning that they cannot be assigned to the same table. That said, similar solutions are achieved for $p = 0.0$, $0.3$, and $0.6$ for various values of $k$, suggesting that the cost of the best solutions found is not unduly affected by the presence of moderate levels of hard constraints. The exception to this pattern, as shown in the figure, is for the smallest achievable values for $k$. Here, the larger number of guest groups per table makes it more likely that combinations of guest groups will be deemed infeasible, reducing the number of possible feasible solutions, and making the presence of a zero-cost solution less likely.

Figure 6.6 now shows the effect that variations in $p$ and $k$ have on the neighbourhood sizes encountered in Stage 2, together with the overall run times of the algorithm. For unconstrained problem instances ($p = 0.0$) all Kempe chains are of size 1, and all pairs of vertices in different colours qualify for a pair swap. Hence the number of distinct moves available for each operator are $n(k-1)$ and approximately $(n(n-n/k))/2$ respectively. However, for more constrained problems (lower $k$'s and/or larger $p$'s), the number of neighbouring solutions becomes far fewer. This means that a smaller number of evaluations need to take place at each iteration of tabu search, resulting in shorter run times. The exception to this pattern is for low values of $k$ using $p = 0.0$, where the larger numbers of guests per table requires more overheads in the calculation of Kempe chains and the cost function, resulting in increased run times.

Finally it is also instructive to consider the proportion of Kempe chains that are seen to be *total* during runs of the tabu search algorithm. Recall from Section 4.1.5 that a Kempe chain KEMPE$(v,i,j)$ is described as total when KEMPE$(v,i,j) = (S_i \cup S_j)$: that is, the graph induced by the set of vertices $S_i \cup S_j$ forms a connected bipartite graph. (Consider, for example, the chain KEMPE$(v_3, 4, 2)$ from Figure 6.3.) It is obvious that interchanging the colours of vertices in a total Kempe chain serves no purpose since this only results in the labels of the two colour classes being swapped,

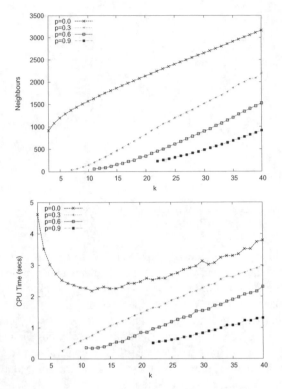

**Fig. 6.6** Average number of neighbouring solutions per iteration of the tabu search algorithm (left); and average run times of the algorithm (right) for the four problem instances using various $k$-values (using a 3.0 GHz Windows 7 PC with 3.87 GB RAM)

thereby having no effect on the objective function. Figure 6.7 shows these proportions for the four considered problem instances. We see that total Kempe chains are more likely to occur with higher values of $p$ (due to the greater connectivity of the graphs), and for lower values of $k$ (because the vertices of the graph are more likely to belong to one of the two colours being considered). Indeed, for $p = 0.9$ and $k = 22$ we see that *all* Kempe chains considered by the algorithm are total, meaning that the neighbourhood operator is ineffective in this case.

## 6.5  Comparison to an IP Model

In this section we now compare the results achieved by our two-stage tabu search algorithm to those of a commercial integer programming (IP) solver. As we saw in Section 3.1.2, one of the advantages of using an IP approach is that, given excess time, we can determine with certainty the optimal solution to a problem instance (or, indeed, whether a feasible solution actually exists). In contrast to our tabu search-

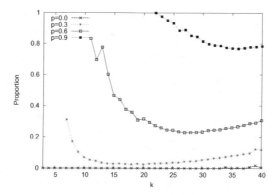

**Fig. 6.7** Proportion of Kempe chains seen to be total for the four problem instances using various $k$-values

based method, the IP solver is therefore able to provide the user with a certificate of optimality and/or infeasibility, at which point it can be halted. Of course, due to the underlying intractability of the WSP these certificates will not always be produced in reasonable time, but given the relatively small problem sizes being considered in this chapter, it is still pertinent to ask how often this is the case, and to also compare the quality of the IP solver's solutions to our tabu search approach under similar time limits.

### 6.5.1 IP Formulation

The WSP can formulated as an IP problem using an extension to the model described the Section 3.1.2 (specifically, Constraints (3.4)–(3.5), and (3.8)–(3.9)). We now outline this model in full. Recall that there are $n$ guest groups that we seek to partition on to $k$ tables. Accordingly, the seating preferences of guests can be expressed using a symmetric matrix $\mathbf{W}_{n \times n}$, where

$$W_{ij} = \begin{cases} \infty & \text{if we require guest groups } i \text{ and } j \text{ to be "definitely apart";} \\ 1 & \text{if we would prefer } i \text{ and } j \text{ to be on different tables ("rather apart");} \\ -1 & \text{if we would prefer } i \text{ and } j \text{ to be on the same table ("rather together");} \\ 0 & \text{otherwise.} \end{cases}$$

(6.5)

As before, we also let $s_i$ define the size of each guest group $i \in \{1, \ldots, n\}$ A solution to the problem can then be represented by the binary matrix $\mathbf{X}_{n \times n}$, where

$$X_{it} = \begin{cases} 1 & \text{if guest group } i \text{ is assigned to table } t, \\ 0 & \text{otherwise,} \end{cases}$$

(6.6)

and the binary vector $\mathbf{Y}_n$ where

$$Y_t = \begin{cases} 1 \text{ if at least one guest group is assigned to table } t, \\ 0 \text{ otherwise.} \end{cases} \qquad (6.7)$$

The following constraints must now be satisfied:

$$\sum_{t=1}^{n} X_{it} = 1 \quad \forall i \in \{1, \ldots, n\} \qquad (6.8)$$

$$X_{it} + X_{jt} \leq Y_t \quad \forall i, j : W_{ij} = \infty, \forall t \in \{1, \ldots, n\} \qquad (6.9)$$

$$X_{it} = 0 \quad \forall i \in \{1, \ldots, n\}, \forall t \in \{i+1, \ldots, n\} \qquad (6.10)$$

$$X_{it} = \sum_{j=t-1}^{i-1} X_{j,t-1} \quad \forall i \in \{2, \ldots, n\}, \forall t \in \{2, \ldots, i-1\} \qquad (6.11)$$

$$\sum_{t=1}^{n} Y_t \leq k \qquad (6.12)$$

Here, Constraint (6.8) ensures that all guest groups are assigned to exactly one table, while Constraint (6.9) stipulates that all "Definitely Apart" constraints are obeyed and that $Y_t = 1$ if and only if there exists a guest group that has been assigned to table $t$. Constraints (6.10) and (6.11) then impose the anti-symmetry constraints (as reviewed in Section 3.1.2), while Constraint (6.12) states that no more than $k$ tables should be used.

As with the tabu search algorithm, the quality of a feasible candidate solution in the IP model is quantified using the sum of the two previously defined objective functions $(f_1 + f_2)$. For the IP model, $f_1$ is rewritten

$$f_1 = \sum_{t=1}^{k} \sum_{i=1}^{n-1} \sum_{j=i+1}^{n} X_{it} X_{jt} (s_i + s_j) W_{ij} \qquad (6.13)$$

in order to cope with the binary matrix method of solution representation; however, it is equivalent in form to Equation (6.3). Similarly, $f_2$ in the IP model is defined in the same manner as Equation (6.4), except that $\tau_i$ is now calculated as $\tau_i = \sum_{j=1}^{n} X_{jt} s_j$. Again, this is equivalent to Equation (6.4).

## 6.5.2 Results

The above IP formulation was implemented and solved using the commercial software FICO Xpress (version 7.5). The experiments of Section 6.4 were repeated using two time limits, five seconds, which is approximately the longest time required by our two-stage tabu search algorithm (see Figure 6.6), and 600 seconds, to gain a broader view of the IP solver's capabilities with this problem.

The results of these trials are summarised in Figure 6.8. The circled lines in the left of the graphs indicate values of $k$ where certificates of infeasibility were produced by the IP solver under the two time limits. As might be expected, these

certificates are produced for a larger range of $k$-values when the longer time limit
is used; however, for $p = 0.3$ and $0.6$ there remain values of $k$ for which feasible
solutions have not been produced (by any algorithm) and where certificates of infeasibility have not been supplied. Thus we are none the wiser as to whether feasible
solutions exist in these particular cases. Also note that certificates of *optimality* were
not provided by the IP solver in any of the trials conducted.

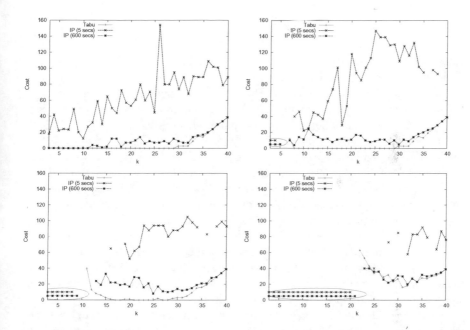

**Fig. 6.8** Comparison of solution costs achieved using the IP solver (using two different time limits)
and our tabu search-based approach for $p = 0.0$, $p = 0.3$, $p = 0.6$, and $p = 0.9$ respectively

Figure 6.8 also reveals that, under the five second limit, the IP approach has
never produced a solution of equal of superior quality to that of our two-stage tabu
search method. In addition, the IP method under this time limit has achieved feasible
solutions in only a subset of cases compared to tabu search (83% of cases). When
the extended run time limit of 600 seconds is applied to the IP solver, this gap in
performance diminishes as we might expect, but similar patterns still emerge. We
see that the tabu search algorithm has produced feasible solutions whenever the
IP approach has, plus for three further cases. In addition, in the 118 cases where
both algorithms have achieved feasibility, tabu search has produced solutions of
superior quality in 91 cases, compared to the IP method's six. Note that these six
instances seem to correspond quite strongly to the runs of tabu search where high
proportions of the Kempe chains are seen to be total, indicating that difficulties are
being experienced when trying to navigate through the space of feasible solutions
with these highly constrained problem instances. However, we must bear in mind

that the IP solver has required more than 400 times the CPU time of the tabu search method to achieve these particular solutions, making it far less suitable for such an online tool.

## 6.6 Chapter Summary and Discussion

In this chapter we have examined the interesting combinatorial problem of constructing seating plans for large parties. As we have seen, the underlying graph colouring properties of this problem allow the design of an effective two-stage heuristic algorithm with much better performance than an equivalent IP formulation, even for fairly small problems. This algorithm might also be used for other situations where we are seeking to divide people into groups, such as birthday parties, gala dinners, team building exercises, and group projects.

The problem formulation considered here (and used with the online tool) is chosen to strike the right balance between being useful to users and being easy to understand. As part of this, we have chosen to allow only three different weights on edges: $-1$, $1$, and $\infty$, corresponding to the constraints "Rather Together", "Rather Apart", and "Definitely Apart" respectively. In practice however, this algorithm could be applied to any set of weights. For example, we might choose to define a threshold value $c$, and then consider any edge $\{u,v\}$ with weight $w_{uv} \geq c$ as a hard constraint, with the remaining edges then being treated as soft constraints.[2] On the other hand, if it is preferable to treat all constraints as soft constraints, we might simply abandon Stage 1 and use the optimisation process of Stage 2 to search through the solution space comprising all $k$-partitions of the guest groups.

Finally, an additional advantage of using a graph colouring-based model is that it can be easily extended to incorporate "table specific" constraints that specify which table each guest group can and cannot be assigned to. To impose such constraints we first need to add $k$ additional vertices to the model, one for each available table. Next, edges of weight $\infty$ then need to be added between each pair of these "table-vertices", thereby forming a clique of size $k$ and ensuring that each table-vertex is assigned to a different colour in any feasible solution. Having introduced these extra vertices we can then add other types of constraints into the model:

- If guest group $v$ is not permitted to sit at table $i$, then an edge of weight $\infty$ can be imposed between vertex $v$ and the $i$th table vertex.
- If a guest group $v$ must be assigned to table $i$, then edges of weight $\infty$ can be imposed between vertex $v$ and all table vertices except the $i$th table vertex.

Note that if our model is to be extended in this way, we will now be associating each subset of guest groups $S_i$ in a solution $\mathcal{S} = \{S_1, \ldots, S_k\}$ with a particular table number $i$. Hence we might also permit tables of different sizes and shapes into the model, perhaps incorporating constraints concerning these factors into the objective

---

[2] That is, the graph $G'$ used in Stages 1 and 2 would comprise edge set $E' = \{\{u,v\} \in E : w_{uv} \geq c\}$.

function. Extensions of this type will be considered in relation to a different problem in the next chapter.

# Chapter 7
# Designing Sports Leagues

In this chapter, our next case study considers the applicability of graph colouring methods for producing *round-robin* tournaments, which are particularly common in sporting competitions. As we will see, the task of producing valid round-robin tournaments for a given number of teams is relatively straightforward, though things can become more complicated when additional constraints are added to the problem. The initial sections of this chapter focus on the problem of producing round-robins in general terms and examine the close relationship between this problem and graph colouring. A detailed real-world case study that makes use of various graph colouring techniques is then presented in Section 7.6.

## 7.1 Problem Background

Round-robin schedules are used in many sports tournaments and leagues across the globe, including the Six Nations Rugby Championships, various European and South American domestic soccer leagues, and the England and Wales County Cricket Championships. Round-robins are schedules involving $t$ teams, where each team is required to play all other teams exactly $l$ times within a fixed number of *rounds*. The most common types are *single* round-robins, where $l = 1$, and *double* round-robins, where $l = 2$. In the latter, teams are typically scheduled to meet once in each other's home venue.

Usually, the number of teams in a round-robin schedule will be even. In cases where $t$ is odd, an extra "dummy team" can be introduced, and teams assigned to play this dummy team will receive a bye in the appropriate part of the schedule.

**Definition 7.1** *Round-robin schedules involving t teams are considered* valid *if each team competes at most once per round. They are also described as* compact *if the number of rounds used is minimal at $l(t-1)$, thus implying $t/2$ matches per round.*

We saw in Chapter 5 that compact round-robin schedules can be constructed for any number of teams $t$ by simply making use of Kirkman's circle method (see

© Springer International Publishing Switzerland 2016
R.M.R. Lewis, *A Guide to Graph Colouring*,
DOI 10.1007/978-3-319-25730-3_7

Theorem 5.10). In addition to this, it is also known that the number of *distinct* round-robin schedules grows exponentially with the number of teams $t$, since this figure is monotonically related to the number of nonisomorphic one-factorisations of the complete graph $K_t$.

**Definition 7.2** *Let $K_t$ be the complete graph with $t$ vertices, where $t$ is even. A one-factor of $K_t$ is a perfect matching. A one-factorisation of $K_t$ is a partition of the edges into $t - 1$ disjoint one-factors.*

An example one-factorisation of $K_6$, comprising five one-factors, is illustrated in Figure 5.7. For $t \in \{2,4,6\}$ there is just one nonisomorphic one-factorisation available. These numbers then rise to 6, 396, 526,915,620, and 1,132,835,421,602,062,347 for $t = 8$ to 14 respectively (Dinitz et al., 1994). Such a growth rate—combined with the fact that many different round-robins can be generated from a particular one-factorisation by relabelling the teams and reordering the rounds—implies that the enumeration of all round-robin schedules will not be possible in reasonable time for nontrivial values of $t$.

In addition to Kirkman's circle method, other algorithms of linear complexity are also available for quickly producing valid compact round-robin schedules. The "greedy round-robin" algorithm, for example, operates by arranging matches into lexicographic order: $\{1,2\}, \{1,3\}, \{1,4\}, \ldots, \{t - 1, t\}$. The first match is then assigned to the first round, and each remaining match is then considered in turn and assigned to the next round where no clash occurs. When the final round is reached, the algorithm loops back to the first round. The solution produced by this method for $t = 8$ teams is given in Figure 7.1. Note that if a double round-robin is required, the second half of the schedule can be produced by simply copying the first half.

Round Matches
$r_1$ :    $\{\{1,2\}, \{3,7\}, \{4,6\}, \{5,8\}\}$
$r_2$ :    $\{\{1,3\}, \{2,8\}, \{4,7\}, \{5,6\}\}$
$r_3$ :    $\{\{1,4\}, \{2,3\}, \{5,7\}, \{6,8\}\}$
$r_4$ :    $\{\{1,5\}, \{2,4\}, \{3,8\}, \{6,7\}\}$
$r_5$ :    $\{\{1,6\}, \{2,5\}, \{3,4\}, \{7,8\}\}$
$r_6$ :    $\{\{1,7\}, \{2,6\}, \{3,5\}, \{4,8\}\}$
$r_7$ :    $\{\{1,8\}, \{2,7\}, \{3,6\}, \{4,5\}\}$

**Fig. 7.1** Single round-robin schedule produced by the greedy round-robin algorithm for $t = 8$ teams

Another linear-complexity method for producing compact round-robin schedules is the "canonical" round-robin algorithm of de Werra (1988). Unlike the circle and greedy methods, this approach focusses on the issue of deciding whether teams should play at home or away. In this case, a "break" is defined as a situation where a team is required to play two home-matches (or away-matches) in consecutive rounds, and it is proven that this method always achieves the minimum number of $t - 2$ breaks. Pseudocode for this method is shown in Figure 7.2. Note that in this case, matches are denoted by an ordered pair with the first and second elements denoting the home- and away-teams respectively. The canonical schedule for $t = 8$ is also given in Figure 7.3.

The task of minimising breaks has also been explored in other research. Trick (2001), Elf et al. (2003), and Miyashiro and Matsui (2006b), for example, have examined the problem of taking an *existing* single round-robin and then assigning home/away values to each of the matches in order to minimise the number of breaks. Miyashiro and Matsui (2005) have also shown that the problem of deciding whether a home/away assignment exists for a particular schedule such that the theoretical minimum of $t - 2$ breaks is achieved is computable in polynomial time. Meanwhile, the inverse of this problem—taking a fixed home/away pattern and then assigning matches consistent with this pattern—has also been studied by various other authors (Russell and Leung, 1994; de Werra, 1988; Nemhauser and Trick, 1998).

One interesting feature of the greedy, circle and canonical methods is that the solutions they produce are isomorphic. For example, a canonical single round-robin schedule for $t$ teams can be transformed into the schedule produced by the greedy round-robin algorithm by simply converting the ordered pairs into unordered pairs and then reordering the rounds. Similarly, a circle schedule can be transformed into a greedy schedule by relabelling the teams using the mapping $t_1 \leftarrow t_t$ and $t_i \leftarrow t_{i-1} \forall i \in \{2, \ldots, t\}$, with the rounds then being reordered (see also (Anderson, 1991)).

| MAKE-CANONICAL-SCHEDULE $(t)$ |
|---|
| (1)  **for** $i \leftarrow 1$ **to** $n - 1$ **do** |
| (2)      **if** $i$ is odd **then** assign $(i, n)$ to round $r_i$ |
| (3)      **else** assign $(n, i)$ to round $r_i$ |
| (4)      **for** $j \leftarrow 1$ **to** $n/2 - 1$ **do** |
| (5)          $x \leftarrow (i + j) \bmod (n - 1)$ |
| (6)          $y \leftarrow (i - j) \bmod (n - 1)$ |
| (7)          **if** $x = 0$ **then** $x \leftarrow n - 1$ |
| (8)          **if** $y = 0$ **then** $y \leftarrow n - 1$ |
| (9)          **if** $i$ is odd **then** assign $(x, y)$ to round $r_i$ |
| (10)         **else** assign $(y, x)$ to round $r_i$ |

**Fig. 7.2** Procedure for producing a canonical single round-robin schedule for $t$ teams, where $t$ is even

## 7.1.1 Further Round-Robin Constraints

Although it is straightforward to construct valid, compact round-robin schedules using constructive methods such as the greedy, circle, and canonical algorithms, in practical circumstances it is often the case that the production of "high-quality" schedules will depend on additional user requirements and constraints. As we have seen, the minimisation of "breaks" is one example of this. Other constraints, how-

Round  Matches
  $r_1$ :   {(1,8), (2,7), (4,5), (6,3)}
  $r_2$ :   {(3,1), (5,6), (7,4), (8,2)}
  $r_3$ :   {(1,5), (3,8), (4,2), (6,7)}
  $r_4$ :   {(2,6), (5,3), (7,1), (8,4)}
  $r_5$ :   {(1,2), (3,7), (5,8), (6,4)}
  $r_6$ :   {(2,3), (4,1), (7,5), (8,6)}
  $r_7$ :   {(1,6), (3,4), (5,2), (7,8)}

**Fig. 7.3** Canonical round-robin schedule for $t = 8$ teams

ever, can include those associated with demands from broadcasters, various econom-
ical and logistical factors, inter-team politics, policing, and the perceived fairness of
the league. In addition, factors such as the type of sport, the level of competition,
and the country (or countries) involved will also play a part in determining what is
considered "high-quality". An impression of the wide range of such constraints and
requirements can be gained by considering the variety of round-robin scheduling
problems that have previously been tackled in the literature, including German, Aus-
trian, and Italian soccer leagues (Bartsch et al., 2006; della Croce and Oliveri, 2006),
New Zealand basketball leagues (Wright, 2006), amateur tennis tournaments (della
Croce et al., 1999), English county cricket fixtures (Wright, 1994), and American
professional ice hockey (Fleurent and Ferland, 1993). A good survey on the wide
range of problems and solution methods for sports scheduling is also provided by
Kendall et al. (2010).

An oft-quoted example of such requirements is the issue of *carryover* in round-
robin schedules. Here, we consider the possibility of a team's performance being
influenced by its opponents in previous rounds. For example, if $t_i$ is known to be a
very strong team whose opponents are often left injured or demoralised, then a team
that plays $t_i$'s opponents in the next round may well be seen to gain an advantage.[1]
In such cases the aim is to therefore produce a round-robin schedule in which the
overall effects of carryover are minimised. This requirement was first considered
by Russell (1980), who proposed a constructive algorithm able to produce provably
optimal schedules in cases where the number of teams $t$ is a power of 2. More
recently, methods for small-sized problems have also been proposed by Trick (2001)
and Henz et al. (2004), who both make use of constraint programming techniques.

Another set of schedule requirements, which has become somewhat of a benchmark over the past decade, is encapsulated in the travelling tournament problem
(TTP). Originally proposed by Easton et al. (2001), in this problem a compact double round-robin schedule is required where teams play each other twice, once in each
other's home venue. The overriding aim is to then minimise the *distances travelled*
by each team. Geographical constraints such as this are particularly relevant in large
countries such as Brazil and the USA where match venues are typically far apart.

---

[1] As an illustration, in Figure 7.1 we see that team 2, for instance, is scheduled to play the opponents
of team 1 from the previous round on five different occasions. This feature also exists for other
teams; thus this schedule actually contains rather a large amount of carryover.

Consequently, when a team is scheduled to attend a succession of away-matches, instead of returning to their home city after each match, the team travels directly to their next away venue. Note that in addition to the basic round-robin scheduling constraints, this problem also contains elements of the travelling salesman problem, as we are interested in scheduling runs of successive away-matches for each team such that they occur in venues that are close to one another. An early solution method proposed for this formulation was proposed by Easton et al. (2003), who used integer- and constraint-programming techniques. Subsequent proposals, however, focused on metaheuristic techniques, and in particular, neighbourhood search-based algorithms. Good examples of these include the simulated annealing approaches of Anagnostopoulos et al. (2006) and Lim et al. (2006), and the local-search approaches of Di Gaspero and Schaerf (2007) and Ribeiro and Urrutia (2007). In the latter example the authors also consider the added restriction that the double round-robin schedule should be "mirrored": that is, if teams $t_i$ and $t_j$ play in $t_i$'s home stadium in round $r \in \{1, \ldots, t-1\}$, then $t_i$ and $t_j$ should necessarily play each other in $t_j$'s home stadium in round $r + (t - 1)$.

## 7.1.2 Chapter Outline

The above paragraphs illustrate that the requirements of sports scheduling problems can be complex and idiosyncratic. In the remainder of this chapter we will examine the ways in which graph colouring concepts can be used to help find solutions to such problems. In the next section we will describe how basic round-robin scheduling problems can be represented as graph colouring problems. In Section 7.3 we then assess the "difficulty" of solving such graphs using our suite of graph colouring algorithms from Chapter 4. Following this, in Section 7.4 we will then discuss ways in which this model can be extended in order to incorporate other types of "hard" (i.e., mandatory) constraint, and in Section 7.5 we discuss various different neighbourhood operators that can be used with this extended model for exploring the space of *feasible* solutions (that is, round-robin solutions that are compact, valid, and also obey any imposed hard constraints). Finally, in Section 7.6 we consider a real-world round-robin scheduling problem from the Welsh Rugby Union and propose two separate algorithms that make use of our proposed algorithmic operators. The performance of these algorithms is then analysed over a number of different problem instances.

## 7.2 Representing Round-Robins as Graph Colouring Problems

Round-robin scheduling problems can be represented as graph colouring problems by considering each individual match as a vertex, with edges then being added between any pair of matches that cannot be scheduled in the same round (i.e., matches featuring a common team). Colours then represent the individual rounds of the schedule, and the task is to colour the graph using $k$ colours, where $k$ represents the number of available rounds. Note that for the remainder of this chapter we only consider the task of producing compact schedules: thus, $k = \chi(G)$ unless otherwise specified.

For the single round-robin, each vertex is associated with an unordered pair $\{t_i, t_j\}$, denoting a match between teams $t_i$ and $t_j$. The number of vertices $n$ in such graphs is thus $\frac{1}{2}t(t-1)$, with $\deg(v) = 2(t-2) \; \forall v \in V$. For a compact schedule, the number of available colours $k = t - 1$. For double round-robins the number of vertices $n = t(t-1)$, $\deg(v) = 4(t-2) + 1 \; \forall v \in V$, and $k = 2(t-1)$, since teams will play each other twice. In this case, each vertex is associated with an ordered pair $(t_i, t_{j \neq i})$, with $t_i$ denoting the home-team and $t_j$ the away-team. An example graph for a double round-robin with $t = 4$ teams is provided in Figure 7.4.

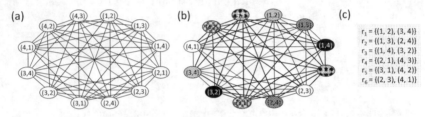

**Fig. 7.4** Graph for a double round-robin problem with $t = 4$ teams (a), an optimal colouring of this graph (b), and the corresponding schedule (c)

Recall from Section 5.2 that the complete graph $K_t$ can also be used to represent a round-robin scheduling problem by associating each vertex with a team and each edge with a match. In such cases, the task is to find a proper edge colouring of $K_t$, with all edges of a particular colour indicating the matches that occur in a particular round. The graphs generated using our methods above are the corresponding *line graphs* of these complete graphs. Of course, in practice it is easy to switch between these two representations. However, the main advantage of using our representation is that it allows the exploitation of previously developed vertex-colouring techniques, as the following sections will demonstrate.

## 7.3 Generating Valid Round-Robin Schedules

Having defined the basic structures of the "round-robin graphs" that we wish to colour, in this section we investigate whether such graphs actually constitute difficult-to-colour problem instances. Note that by $k$-colouring such graphs we are doing nothing more than producing valid, compact round-robin schedules which, as we have mentioned, can be easily constructed using the circle, greedy, and canonical algorithms. However, there are a number of reasons why solving these problems from the perspective of vertex colouring is worthwhile.

- Because of the structured, deterministic way in which the circle, greedy and canonical methods operate, their range of output will only represent a very small part of the space of all valid round-robin schedules.
- The schedules that are produced by the circle, greedy and canonical methods also occupy very *particular* parts of the solution space. For example, Miyashiro and Matsui (2006a) have conjectured that the circle method produces schedules in which the amount of carryover is actually maximised.
- As noted earlier, the solutions produced via the greedy and circle and canonical methods are in fact isomorphic. Moreover, the specific structures present in these isomorphic schedules are often seen to have adverse effects when applying neighbourhood search operators, as we will see in Section 7.5.
- Finally, we are also able to modify the graph colouring model to incorporate additional real-world constraints, as shown in Section 7.4.

By using graph colouring methods, particularly those that are stochastic in nature, the hope is that we therefore have a more robust and less biased mechanism for producing round-robin schedules, allowing a larger range of structurally distinct schedules to be sampled. This is especially useful in the application of metaheuristics, where the production of random initial solutions is often desirable.

Figure 7.5 summarises the results of experiments using single and double round-robins of up to $t = 60$ teams. 50 runs of the backtracking and hybrid evolutionary algorithms were executed in each case using a computation limit of $5 \times 10^{11}$ constraint checks as before. The success rates in these figures gives the percentage of these runs where optimal colourings (compact valid round-robins) were produced. It is obvious from these figures that the HEA is very successful here, featuring 100% success rates across all instances. Indeed, no more than of 0.006% of the computation limit on average was required for any of the values of $t$ tested. On the other hand, the backtracking approach experiences more difficulty, with success rates dropping considerably for larger values of $t$. That said, when the algorithm *does* produce optimal solutions it does so quickly, indicating that solutions are either found early in the search tree, or not at all.[2] The success of the HEA with these instances is also reinforced by the fact that its solutions are very diverse, as illustrated in the figure.

## 7.4 Extending the Graph Colouring Model

Another advantage of transforming the task of round robin construction into a type of graph colouring problem is that we can easily extend the model to incorporate

---

[2] On this point, Lewis and Thompson (2010) have also found that much better results for the backtracking algorithm on these particular graphs can be achieved by restricting the algorithm to only inspect one additional branch from each node of the search tree. The source code available for this algorithm can easily be modified to allow this.

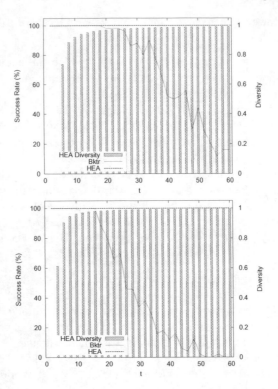

**Fig. 7.5** Success rates of the Backtracking and HEA algorithms for finding optimal colourings with, respectively, single round-robin graphs for $t = 2, \ldots, 60$ ($n = 1, \ldots, 1,770$), and double round-robin graphs ($n = 2, \ldots, 3,540$). All figures are averaged across 50 runs. The bars show the diversity of solutions produced by the HEA across the 50 runs, calculated using Equation (4.9)

other types of sports scheduling constraints. In this section, we specifically consider the imposition of *round-specific* constraints, which specify the rounds that matches can and cannot be assigned to. In many practical cases, round-specific constraints will be a type of hard constraint—that is, they will be mandatory in their satisfaction—and candidate solutions that violate such constraints will be considered infeasible. Encoding such constraints directly into the graph colouring model thus allows us to attach an importance to these constraints that is equal to the basic round-robin constraints themselves.

To impose round-specific constraints we follow the method seen in Section 5.7 for the list colouring problem. First, $k$ extra vertices are added to the model, one for each available round. Next, edges are then added between all pairs of these "round-vertices" to form a clique of size $k$, ensuring that each round-vertex will be assigned to a different colour in any feasible solution. Having introduced these extra vertices, a variety of different round-specific constraints can then be introduced:

Match Unavailability Constraints:   Often it will be impossible to assign a match
to a particular round, perhaps because the venue of the match is being used for
another event, or because league rules state that the associated teams should not
play each other in specific rounds in the league. Such constraints are introduced
by adding an edge between the relevant match-vertex and round-vertex.

Preassignment Constraints:   In some cases, a match will need to be assigned to
a specific round $r_i$, e.g., to increase viewing figures and/or revenue. For such
constraints, edges are added between the appropriate match-vertex and all round-
vertices except for the round-vertex corresponding to $r_i$.

Concurrent Match Constraints:   In some cases it might also be undesirable for two
matches $(t_i, t_j)$ and $(t_l, t_m)$ to occur in the same round. For example, the two home
teams $t_i$ and $t_l$ may share a stadium, or perhaps it is forbidden for rival fans of
certain teams to visit the same city on the same day. Such constraints can be
introduced by assigning an edge between the appropriate pairs of match-vertices.

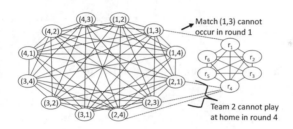

**Fig. 7.6** Method of extending the graph colouring model to incorporate round-specific constraints.
"Match-vertices" appear on the left; "round-vertices" on the right

Figure 7.6 gives two examples of how we can impose such constraints. It is obvious that any feasible $k$-coloured solution for such graphs will constitute a valid
compact round-robin schedule that obeys the imposed round-specific constraints.
As we will see in Section 7.5, incorporating constraints in this fashion also allows
us to apply neighbourhood operators stemming from the underlying graph colouring model that ensure that these extra constraints are never reviolated. Note that an
alternative strategy for coping with hard constraints such as these is to *allow* their
violation within a schedule, but to then penalise their occurrence via a cost function.
Anagnostopoulos et al. (2006), for example, use a strategy whereby the space of all
valid compact round-robins is explored, with a cost function then being used that
reflects the number of hard *and soft* constraint violations. Weights are then used to
place a higher penalty on violations of the hard constraints, and it is hoped that by
using such weights the search will eventually move into areas of the solution space
where no hard constraint violations occur. The choice of which strategy to employ
will depend largely on practical requirements.

To investigate the effects that the imposition of round-specific constraints has
on the difficulty of the underlying graph colouring problem, double round-robin

graphs were generated with varying numbers of match unavailability constraints. Specifically, these constraints were added by considering each match-vertex/round-vertex pair in turn, and adding edges between them with probability $p$. This means, for example, that if $p = 0.5$, each match can only be assigned to approximately half of the available rounds. Graphs were also generated in two ways: one where $k = \chi(G)$ was ensured (by making reference to a pregenerated valid round-robin), and one where this matter was ignored, possibly resulting in graphs for which $\chi(G) > k$.

Note that by adding edges in this binomially distributed manner, the expected degrees of each vertex can be calculated in the following way. Let $V_1$ define the set of match-vertices and $V_2$ the set of round-vertices, and let $v \in V_1$ and $u \in V_2$. Then:

$$
\begin{aligned}
\mathbb{E}(\deg(v)) &= 4(t-2)+1+p \times |V_2| \quad \forall v \in V_1, \text{ and} \\
\mathbb{E}(\deg(u)) &= 2(t-1)-1+p \times |V_1| \quad \forall u \in V_2.
\end{aligned}
\tag{7.1}
$$

The expected variance in degree across all vertices $V = V_1 \cup V_2$, where $n = |V|$, is thus approximated as

$$
\frac{|V_1| \times \mathbb{E}(\deg(v))^2 + |V_2| \times \mathbb{E}(\deg(u))^2}{n} - \left( \frac{|V_1| \times \mathbb{E}(\deg(v)) + |V_2| \times \mathbb{E}(\deg(u))}{n} \right)^2.
\tag{7.2}
$$

The effect that $p$ has on the overall degree coefficient of variation (CV) of these graphs is demonstrated in Figure 7.7. As $p$ is increased from zero, $\mathbb{E}(\deg(v))$ and $\mathbb{E}(\deg(u))$ initially become more alike, resulting in a slight drop in the CV. However, as $p$ is increased further, $\mathbb{E}(\deg(u))$ rises more quickly than $\mathbb{E}(\deg(v))$, resulting in large increases to the CV.

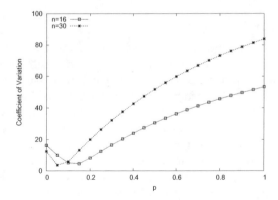

**Fig. 7.7** Effect of varying $p$ on the degree coefficient of variation with double round-robin graphs of size $t = 16$ and 30

The consequences of these specific characteristics help to explain the performance of our six graph colouring algorithms across a large number of instances, as shown in Figure 7.8. As with the results from Chapter 4, the quality of solution

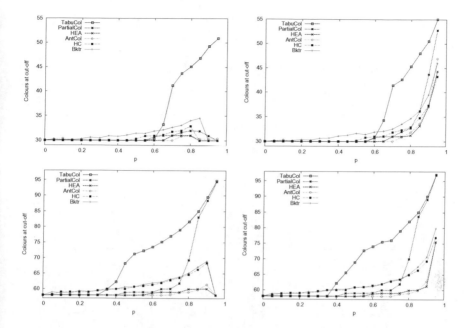

**Fig. 7.8** Mean quality of solutions achieved with double round-robin graphs using (respectively): $t = 16$, $(n = 270, k = 30)$ with $\chi(G) = 30$; $t = 16$ with $\chi(G) \geq 30$; $t = 30$, $(n = 928, k = 58)$ with $\chi(G) = 58$; and $t = 30$ with $\chi(G) \geq 58$. All points are the average of 25 runs on 25 graphs

achieved by TABUCOL and PARTIALCOL is observed to be substantially worse than that of the other approaches when $p$, and therefore the degree CV, is high. In particular, TABUCOL shows very disappointing performance, providing the worst-quality results for both sizes of graph in all cases where the CV is $\geq 40\%$.

In contrast, some of the best performance across the instances is once again due to the HEA. Surprisingly, ANTCOL also performs well here, with no significant difference being observed in the mean results of the HEA and ANTCOL algorithms across the set. The reasons for the improved performance of ANTCOL, particularly with denser graphs, seems due to two factors: (a) the higher degrees of the vertices in the graphs, and (b) the high variance in degrees. In ANTCOL's BUILDSOLUTION procedure (Section 4.1.4) the first factor naturally increases the influence of the heuristic value $\eta$ in Equation (4.4), while the second allows a greater discrimination between vertices. In these cases it seems that a favourable balance between heuristic and pheromone information is being struck, allowing ANTCOL's global operator to effectively contribute to the search.

Note, however, that if we examine individual values of $p$, the picture becomes more complicated. Figure 7.9, for example, shows run profiles of ANTCOL and HEA on four sets of large, highly constrained round-robin graphs. For comparison, the variant HEA* is also included here, which features a local-search iteration limit equal to ANTCOL's (i.e., $I$ reduced from $16n$ to $2n$). For $p = 0.8$, we see

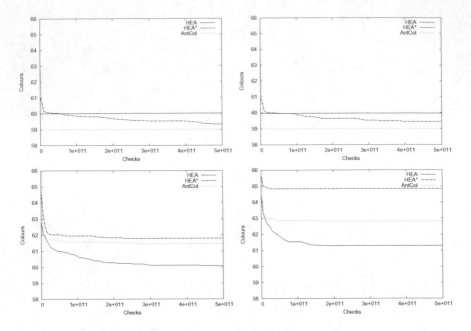

**Fig. 7.9** Run profiles for double round-robins with $t = 30$ ($n = 928$) using respectively: $p = 0.8$, $\chi(G) = 58$; $p = 0.8$, $\chi(G) \geq 58$; $p = 0.9$, $\chi(G) = 58$; and $p = 0.9$, $\chi(G) \geq 58$. HEA* denotes the HEA algorithm with a reduced local search limit of $I = 2n$

that ANTCOL quickly produces the best observed results, with HEA consistently requiring an additional colour. However, reducing the local search element of the HEA—thus placing more emphasis on global search—seems to improve performance, though the results of HEA* are still inferior to those of ANTCOL. On the other hand, for $p = 0.9$ the picture is reversed, with HEA (using $I = 16n$) clearly producing the best results, suggesting that increased amounts of local search are beneficial in this case.

Despite these complications however, the results of Figure 7.9 allow us to conclude that methods such as ANTCOL and the HEA provide useful mechanisms for producing feasible compact round-robin schedules, even in the presence of high levels of additional constraints.

## 7.5  Exploring the Space of Round-Robins

Upon production of a valid round-robin schedule, we may now choose to apply one or more neighbourhood operators to try and eliminate occurrences of any remaining soft constraint violations. Table 7.1 lists a number of neighbourhood operators that have been proposed for round-robin schedules, mostly for use with the travelling tournament problem (Anagnostopoulos et al., 2006; Di Gaspero and Schaerf, 2006; Ribeiro and Urrutia, 2007). Note that the information given in this table applies to double round-robins, so the number of rounds $k = 2(t-1)$; however, simple adjustments can be made for other cases.

A point to note about these operators is that while they all preserve the validity of a round-robin schedule, they will not be useful in all circumstances. For example, applications of $N_1$, $N_2$, and $N_3$ will not affect the amount of carryover in a schedule. Also, while applications of $N_2$ can change the home/away patterns of individual teams, they cannot alter the *total* number of breaks in a schedule. Finally, perhaps the most salient point from our perspective is that if extra hard constraints are being considered, such as the round-specific constraints listed in Section 7.4, then the application of such operators may lead to schedules that, while valid, are not necessarily feasible.

**Table 7.1** Description of various neighbourhood operators that preserve the validity and compactness of double round-robin schedules. "Move Size" refers to the number of matches (vertices in the graph colouring model) that are affected by the application of these operators

| Description | Move Size |
|---|---|
| $N_1$ Select two teams, $t_i \neq t_j$, and swap the rounds of vertices $(t_i,t_j)$ and $(t_j,t_i)$. | 2 |
| $N_2$ Select two teams, $t_i \neq t_j$, and swap their opponents in all rounds. | $2(k-2)$ |
| $N_3$ Select two teams, $t_i \neq t_j$, and swap all occurrences of $t_i$ to $t_j$ and all occurrences of $t_j$ to $t_i$. | $2k-2$ |
| $N_4$ Select two rounds, $r_i \neq r_j$ and swap their contents. | $t$ |
| $N_5$ Select a match and move it to a new round. Repair the schedule using an ejection chain repair procedure. | Variable |

Pursuing the relationship with graph colouring, a promising strategy for exploring the space of round-robin schedules is again presented by the Kempe chain interchange operator (see Definition 3.1). Of course, because this operator is known to preserve the feasibility of a graph colouring solution, it is applicable to both the basic and extended versions of our graph colouring model. On the other hand, the pair-swap operator (Definition 3.2) is not suitable here because swapping the colours of two nonadjacent vertices (i.e., swapping the rounds of a pair of matches with no common team) will not retain the feasibility of a solution.

Recall that the number of vertices affected by a Kempe chain interchange can vary. For basic (non-extended) round-robin colouring problems involving $t$ teams, the largest possible move involves $t$ vertices (i.e., two colours, with $t/2$ vertices in each). In Figure 7.10 we illustrate what we have found to be typical-shaped distributions of the differently sized Kempe chains with single and double round-robins. These examples were gained by generating initial solutions with our graph colouring algorithms and then performing random walks of $10^6$ neighbourhood moves. We see that in the case of double round-robins, the smallest moves involve exactly two vertices, which only occurs when a chain is formed containing the complementary

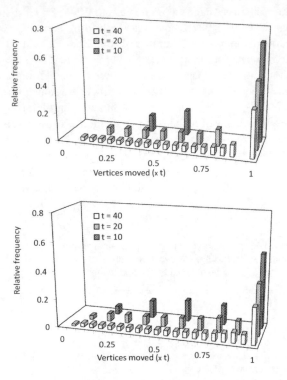

**Fig. 7.10** Distribution of differently sized Kempe chains for $t = 10, 20,$ and $40$ for SRRs and DRRs

match-vertices $(t_i, t_j)$ and $(t_j, t_i)$. In this case the Kempe move is equivalent to the operator $N_1$ (Table 7.1) and it occurs with a probability $\frac{1}{k-1}$ (obviously moves of size 2 do not occur with single round-robins because a match does not have a corresponding reverse fixture with which to be swapped). Meanwhile, the most probable move in both cases is a total Kempe chain interchange (i.e., involving all vertices in the two associated colours). Moves of this size are equivalent to a corresponding move in $N_4$ and occur when all vertices in the two colours form a connected component. Such moves appear to be quite probable due to the relatively high edge densities of the graphs. Importantly however, we see that for larger values of $t$ the majority of moves are of sizes between these two extremes, resulting in moves that are beyond those achievable with neighbourhood operators $N_1$ and $N_4$.

We also performed this same set of experiments by performing random walks from schedules generated by the circle, greedy and canonical algorithms. However, due to the structured way in which these methods go about constructing a schedule, for many different values of $t$ single round-robins are produced in which *all* applications of the Kempe chain interchange operator are of size $t$.[3] For DRRs similar

---

[3] Such solutions are usually termed *perfect* one-factorisations. Specifically, this occurs with 33.6% of $t$ values between 2 and 1,000, with the rate of occurrence remaining fairly consistent between

observations were also made, though in this case moves of size 2 were also possible with the remaining proportion of $1 - \frac{1}{k-1}$ moves being of size $t$. Clearly, such features are undesirable as they do not allow the Kempe chain interchange operator to produce moves beyond what can already be achieved using $N_1$ and $N_4$, limiting the number of solutions accessible via the operator. We should note, however, that we found that this problem could be circumnavigated in some cases by applying neighbourhood operator $N_5$ from Table 7.1 to the solution. It seems that, unlike the other operators detailed in this table, $N_5$ has the potential of breaking up the structural properties of these solutions, allowing the Kempe chain distributions to assume their more "natural" shapes as seen in Figure 7.10. However, we still found cases where this situation was not remedied.[4]

One of the main reasons why an analysis of move sizes is relevant here is because of the effects that the size of a move can have on the cost of a solution at different stages of the optimisation process. On the one hand, "large" moves can facilitate the exploration of wide expanses of the solution space and can provide useful mechanisms for escaping local optima. On the other hand, when relatively good candidate solutions are being considered, large moves will also be disruptive, usually worsening the quality of a solution as opposed to improving it. These effects are demonstrated in Figure 7.11 where we illustrate the relationship between the size of a move and the resultant change in an arbitrary cost function. In the left graph, the Kempe chain interchange operator has been repeatedly applied to a solution that was randomly produced by one of our graph colouring algorithms. Note that larger moves here tend to give rise to greater variance in cost, but that many moves lead to improvements. In contrast, on the right-hand side the effects of the Kempe chain interchange operator on a relatively "good" solution (which has a cost approximately quarter of the previous one) are demonstrated. Here, larger moves again feature a larger variance in cost, but we also witness a statistically significant medium positive correlation ($r = 0.46$), demonstrating that larger moves tend to be associated with larger decreases in solution quality. Di Gaspero and Schaerf (2007) have also noted the latter phenomenon (albeit with different neighbourhoods and a different cost function) and have suggested a modification to their neighbourhood search algorithm whereby any move above a specific size is automatically rejected with no cost evaluation taking place. Because such moves lead to a degradation in quality and will therefore be rejected in the majority of cases, they find that their algorithm's performance over time is increased by skipping these mostly unnecessary evaluations. On the flip side, of course, such a strategy also eliminates the possibly of "larger" moves occurring which could diversify the search in a useful way, though this is not seen to be an issue in their work.

---

these bounds. For values of $t < 50$ this occurs with SRRs for $t = 8, 12, 14, 18, 20, 24, 30, 32, 38, 42, 44,$ and $48$.

[4] Specifically for SRRs with $t = 12, 14, 20, 30,$ and $38$.

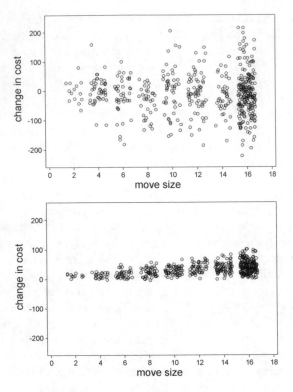

**Fig. 7.11** Demonstrating how Kempe chain interchanges of different sizes influence the change in cost of a randomly generated solution (left) and a "good" solution (right). In both cases a double round-robin with $t = 16$ teams was considered using cost function $c_2$ defined in Section 7.6.1 (negative changes thus reflect an improvement)

## 7.6 Case Study: Welsh Premiership Rugby

In this section we now present a real-world application of the graph colouring-based techniques introduced in this chapter. This problem was provided to us by the Welsh Rugby Union (WRU), based at the Millennium Stadium in Cardiff, which is the governing body for all rugby competitions, national and international, in Wales. The particular problem that we are concerned with is the *Principality Premiership* league, which is the highest-level domestic league in the country.

The Principality Premiership problem involves $t$ teams playing in a compact double round-robin tournament. A number of round-specific (hard) constraints are also stipulated, all of which are mandatory in their satisfaction.

Hard Constraint A:    Some pairs of teams in the league share a home stadium. Therefore when one of these teams plays at home, the other team must play away.

Hard Constraint B:    Some teams in the league also share their stadia with teams from other leagues and sports. These stadia are therefore unavailable in certain rounds. (In practice, the other sports teams using these venues have their matches scheduled before the Principality Premiership teams, and so unavailable rounds are known in advance.)

Hard Constraint C:    Matches involving regional rivals (so-called "derby matches") need to be preassigned to two specific rounds in the league, corresponding to those falling on the Christmas and Easter weekends.

The league administrators also specify two soft constraints. First, they express a preference for keeping reverse fixtures (i.e., matches $(t_i, t_j)$ and $(t_j, t_i)$) at least five rounds apart and, if possible, for reverse fixtures to appear in opposite "halves" of the schedule (they do not consider the stricter requirement of "mirroring" to be important, however). Second, they also express a need for all teams to have good home/away patterns, which means avoiding breaks wherever possible.

## 7.6.1 Solution Methods

In this section we describe two algorithms for this scheduling problem. Both of these use the strategy of first producing a feasible solution, followed by a period of optimisation via neighbourhood search in which feasibility (i.e., validity, compactness, and adherence to all hard constraints) is maintained. Specific details of these methods together with a comparison are given in the next three subsections.

In both cases initial feasible solutions are produced by encoding all of the hard constraints using the extended graph colouring model from Section 7.4, with one of our graph colouring algorithms then being applied. For Hard Constraint A, if a pair of teams $t_i$ and $t_j$ is specified as sharing a stadium then edges are simply added between all match-vertices corresponding to home matches of these teams. For Hard Constraint B, if a venue is specified as unavailable in a particular round, then edges are added between all match-vertices denoting home matches of the venue's team(s) and the associated round-vertex. Finally for Hard Constraint C, edges are also added between the vertices corresponding to derby matches and all round-vertices except those representing derby weekends.

Details of the specific problem instance faced at the WRU are given in bold in Table 7.2. To aid our analysis we also generated (artificially) a further nine instances of comparable size and difficulty, details of which are also given in the table.[5] As it turned out, we found that it was quite straightforward to find a feasible solution to the WRU problem using the graph colouring algorithms from Chapter 4. For the purposes of our experiments we therefore ensured that all artificially generated problems also feature at least one feasible solution. However, the minimum number of soft-constraint violations achievable in these problems is not known.

---

[5] These instances can be downloaded from www.rhydlewis.eu/resources/PrincipalityPremProbs.zip.

**Table 7.2** Summary of the sports scheduling problem instances used. The entry in bold refers to the real-world WRU problem. These problems can be downloaded from www.rhydlewis.eu/resources/PrincipalityPremProbs.zip

| #  | Teams $t$ | Vertices $n$ | Graph Density | A[a] | B[b] | C[c] |
|----|-----------|--------------|---------------|------|--------------------------|------|
| 1  | 12        | 154          | 0.268         | 0    | 2 {5, 5}                 | 3    |
| 2  | 12        | 154          | 0.292         | 1    | 3 {6, 8, 10}             | 6    |
| 3  | 12        | 154          | 0.308         | 2    | 4 {3, 6, 8, 10}          | 6    |
| 4  | 14        | 208          | 0.236         | 0    | 2 {4, 5}                 | 4    |
| **5**  | **14**        | **208**          | **0.260**         | **1**    | **3 {8, 10, 10}**             | **7**    |
| 6  | 14        | 208          | 0.271         | 2    | 5 {3, 6, 8, 10, 10}      | 7    |
| 7  | 16        | 270          | 0.219         | 1    | 3 {4, 5, 6}              | 5    |
| 8  | 16        | 270          | 0.237         | 2    | 5 {3, 6, 8, 10, 10}      | 8    |
| 9  | 18        | 340          | 0.194         | 1    | 3 {4, 5, 6}              | 6    |
| 10 | 18        | 340          | 0.212         | 2    | 6 {4, 5, 6, 7, 10, 10}   | 9    |

[a] Number of pairs of teams sharing a stadium

[b] Number of teams sharing a stadium with teams from another league/sport. The number of match-unavailability constraints for each of these teams is given in { }'s.

[c] Number of local derby pairings

The soft constraints of this problem are captured in two cost functions, both of which need to be minimised:

Spread Cost ($c_1$): Here, a penalty of 1 is added each time a match $(t_i, t_j)$ and its return fixture $(t_j, t_i)$ are scheduled in rounds $r_p$ and $r_q$, such that $|r_p - r_q| \leq 5$. In addition, a penalty of $\frac{1}{1/2 \times t(t-1)}$ is also added each time matches $(t_i, t_j)$ and $(t_j, t_i)$ are scheduled to occur in the same half of the schedule.

Break Cost ($c_2$): Here, the home/away pattern of each team is analysed in turn and penalties of $b^l$ are incurred for each occurrence of $l$ consecutive breaks. In other words, if a team is required to play two home-matches (or away-matches) in succession, this is considered as one break and incurs a penalty of $b^1$. If a team has three consecutive home-matches (or away-matches), this is considered as two consecutive breaks and results in a penalty of $b^2$ being added, and so on.

The term $\frac{1}{1/2 \times t(t-1)}$ is used as part of $c_1$ to ensure that the total penalty due to match pairs occurring in the same half is never greater than 1, thus placing a greater emphasis on keeping matches and their return fixtures at least five rounds apart. The penalty unit of $b^l$ in cost function $c_2$ is also used to help discourage long breaks from occurring in the schedule. In our case, we use $b = 2$: thus a penalty of 2 is incurred for single breaks, 4 for double breaks, 8 for triple breaks, and so on.

It is notable that since the cost functions $c_1$ and $c_2$ measure different characteristics, use different penalty units, and feature different growth rates, they are in some sense incommensurable. For this reason it is appropriate to use the concept of *dominance* in order to distinguish between solutions. This is defined as follows:

**Definition 7.3** Let $\mathcal{S}_1$ and $\mathcal{S}_2$ be two feasible solutions. $\mathcal{S}_1$ is said to dominate $\mathcal{S}_2$ if and only if:

- $c_1(\mathcal{S}_1) \leq c_1(\mathcal{S}_2)$ and $c_2(\mathcal{S}_1) < c_2(\mathcal{S}_2)$; or
- $c_1(\mathcal{S}_1) < c_1(\mathcal{S}_2)$ and $c_2(\mathcal{S}_1) \leq c_2(\mathcal{S}_2)$.

In this definition it is assumed that both cost functions are being minimised. Note that this definition can also be extended to more than two cost functions if required.

If $S_1$ does not dominate $S_2$, and $S_2$ does not dominate $S_1$, then $S_1$ and $S_2$ are said to be *incomparable*. The output to both algorithms is then a list $L$ of mutually incomparable solutions that are not dominated by any other solutions encountered during the search.

Note that the concept of dominance is commonly used in the field of multiobjective optimisation where, in addition to being incommensurable, cost functions are often in conflict with one another (that is, an improvement in one cost will tend to invoke the worsening of another). It is unclear whether the two cost functions used here are necessarily in conflict, however.

### 7.6.1.1 The Multi-stage Approach

Our first method for the WRU problem operates in a series of stages, with each stage being concerned with minimising just one of the cost functions. A description of this method is given in Figure 7.12.

| **Multi-stage Algorithm ($S$)** |
| --- |
| (1) **while** (**not** stopping condition) **do** |
| (2)      Apply perturbation to $S$ |
| (3)      Reduce cost $c_1$ in $S$ via random descent |
| (4)      Reduce cost $c_2$ in $S$ via simulated annealing |
| (5)      Update list $L$ of non-dominated solutions |

**Fig. 7.12** The Multi-stage Algorithm. The procedure takes as input a feasible solution $S$ provided by a graph colouring algorithm

As shown, the algorithm starts by taking an arbitrary feasible solution $S$ produced by a graph colouring method. This solution is then "perturbed" by performing a series of randomly selected Kempe chain interchanges, paying no heed to either cost function. Next, a random descent procedure is applied that attempts to make reductions to the spread cost $c_1$. This is done by repeatedly selecting a random Kempe chain at each iteration and performing the interchange only if the resultant spread cost is less than or equal to the current spread cost. On completion of the random descent procedure, attempts are then made to reduce the break cost $c_2$ of the current schedule *without increasing the current spread cost*. This is achieved using a phase of simulated annealing with a restricted neighbourhood operator where only matches and their reverse fixtures (i.e., $(t_i, t_j)$ and $(t_j, t_i)$) can be swapped. Note that the latter moves can, on occasion, violate some of the additional hard constraints of this problem, and so in these cases such moves are rejected automatically. Also note that moves in this restricted neighbourhood do not alter the spread cost of the schedule and therefore do not undo any of the work carried out in the previous random descent stage. On completion, the best solution $S^*$ found during this round of simulated annealing is used to update $L$. Specifically, if $S^*$ is seen to dominate any solutions in $L$, then these solutions are removed from $L$ and $S^*$ is added to $L$. The entire process is then repeated.

Our choice of random descent for reducing $c_1$ arises simply because in initial experiments we observed that, in isolation, the associated soft constraints seemed quite easy to satisfy. Thus a simple descent procedure seems effective for making quick and significant gains in quality (for all instances spread costs of less than 1, and often 0, were nearly always achieved within our imposed cut-off point of 10,000 evaluations). In addition to this we also noticed that only short execution times were needed for the simulated annealing stage due to the relatively small solution space resulting from the restricted neighbourhood operator, which meant that the search would tend to converge quite quickly at a local optimum. In preliminary experiments we also found that if we lengthened the simulated annealing process by allowing the temperature variable to be reset (thus allowing the search to escape these optima), then the very same optimum would be achieved after another period of search, perhaps suggesting that the convergence points in these searches are the true optima in these particular spaces.[6]

Finally, our use of a perturbation operator in the multi-stage algorithm is intended to encourage a diversification in the search. In this case a balance needs to be struck by applying enough changes to the current solution to cause the search to enter a different part of the solution space, but not applying too many changes so that the operator becomes nothing more than a random restart mechanism. In our case, we chose to simply apply the Kempe chain operator five times in succession, which proved sufficient for our purposes.

### 7.6.1.2  A Multiobjective Optimisation Approach

In contrast to the multi-stage approach, our second method attempts to eliminate violations of both types of soft constraint simultaneously. This is done by combining both cost functions into a single weighted objective function $f(S) = w_1 \times c_1(S) + w_2 \times c_2(S)$, used in conjunction with the Kempe chain neighbourhood operator.

An obvious issue with the objective function $f$ is that suitable values need to be assigned to the weights. Such assignments can, of course, have large effects on the performance of an algorithm, but they are not always easy to determine as they depend on many factors such as the size and type of problem instance, the nature of the individual cost functions, the user requirements, and the amount of available run time. To deal with this issue, we adopt a multiobjective optimisation technique of Petrovic and Bykov (2003). The strategy of this approach is to alter weights dynamically during the search based on the quality of solutions found so far, thus directing the search into specific regions of the solution space. This is achieved by

---

[6] In all cases we used an initial temperature $t = 20$, a cooling rate of $\alpha = 0.99$, and $z = \frac{n}{2}$ (refer to the simulated annealing algorithm in Figure 3.5). The annealing process ended when no move was accepted for 20 successive temperatures. Such parameters were decided upon in preliminary testing and were not seen to be critical in dictating algorithm performance.

providing two *reference costs* to the algorithm, $x_1$ and $x_2$. Using these values, we can then imagine a reference point $(x_1, x_2)$ being plotted in a two-dimensional Cartesian space, with a straight *reference line* then being drawn from the origin $(0,0)$ and through the reference point (see Figure 7.14). During the search, all solutions encountered are then also represented as points in this Cartesian space and, at each iteration, the weights are adjusted automatically to encourage the search to move towards the origin while remaining close to the reference line. It is hoped that eventually solutions will be produced that feature costs less than the original reference costs.

---

**Multiobjective Algorithm** $(\mathcal{S}, \Delta w)$

---

(1) Set reference costs $x_1$ and $x_2$
(2) Set initial weights using $w_i = \frac{c_i(\mathcal{S})}{x_i}$ for $i \in \{1,2\}$
(3) Calculate weighted cost of solution, $f(\mathcal{S}) = w_1 \times c_1(\mathcal{S}) + w_2 \times c_2(\mathcal{S})$
(4) $B \leftarrow f(S)$
(5) **while (not** stopping condition) **do**
(6)      Form new solution $\mathcal{S}'$ by applying a Kempe chain interchange to $\mathcal{S}$
(7)      **if** $(f(\mathcal{S}') \leq f(\mathcal{S})$ **or** $(f(\mathcal{S}') \leq B)$ **then**
(8)          $\mathcal{S} \leftarrow \mathcal{S}'$
(9)          Update list $L$ of non-dominated solutions using $\mathcal{S}$
(10)     Find $i$ corresponding to $\max_{i \in \{1,2\}} \left\{ \frac{c_1(\mathcal{S})}{x_1}, \frac{c_2(\mathcal{S})}{x_2} \right\}$
(11)     Increase weight $w_i \leftarrow w_i(1 + \Delta w)$

---

**Fig. 7.13** Multiobjective algorithm with variable weights (Petrovic and Bykov, 2003). In all reported experiments, a setting of $\Delta w = 10^{-6}$ was used. The input $\mathcal{S}$ is a feasible solution provided by a suitable graph colouring algorithm

A pseudocode description of this approach is given in Figure 7.13. Note that the weight update mechanism used here (line (11)) means that weights are gradually increased during the run. Since, according to line (7), changes to solutions are only permitted if (a) they improve the cost, or (b) if the weighted cost is kept below a constant $B$, this implies that worsening moves become increasingly less likely during execution. Thus the search process is similar in nature to simulated annealing.

### 7.6.1.3 Experimental Analysis

To compare the performance of the two approaches, we performed 100 runs of each algorithm on each of the ten problem instances shown in Table 7.2. In all cases a cut-off point of 20,000,000 evaluations (of either cost function) was used, resulting in run times of approximately five to ten minutes.[7]

Note that unlike with the multi-stage approach, reference costs $x_1$ and $x_2$ need to be supplied to the multiobjective approach. In our case we chose suitable reference costs by performing one run using the multi-stage approach. We then used the costs

---

[7] Using a 3.0 GHz Windows 7 PC with 3.87 GB RAM.

of the resultant solution as the reference costs for the multiobjective algorithm. In runs where more than one non-dominated solution was produced, solutions in $L$ were sorted according to their costs, and the costs of the median solution were used. Note that because problem instance #5 is based on real-world data, in this case we were also able to use the costs of the (manually produced) league schedule used in the competition. However, runs using the reference costs from both sources turned out to be similar in practice, and so we only report results achieved using the former method here.

We now examine the general behaviour of each method. Figure 7.14 displays the costs of a sample of solutions encountered in an example run with each algorithm. From the initial solution at point $(22.4, 686)$ we see that the multi-stage approach quickly finds solutions with very low values of cost $c_1$. It then spends the remainder of the run attempting to make reductions to $c_2$, which is the cause of the bunching of the points in the left of the graph. Due to the position of the reference line (the dotted line in the figure), the multiobjective approach follows a similar pattern to this, though initial progress is considerably slower. However, as the search nears the reference line, the algorithm then considers $c_1$ to have dropped to an appropriate level and so weight $w_2$ (used in evaluation function $f$ (Figure 7.13)) begins to increase such that reductions to cost $c_2$ are also sought. In this latter stage, unlike in the multi-stage approach, improvements to both cost functions are being made simultaneously. As both algorithms' searches converge, we see from the projection (inset) that the multiobjective approach produces the best solutions in this case.

A summary of the results over all ten test problems is provided in Table 7.3. The "best" column here shows the costs of the best solution found across all runs of both algorithms, which we use for comparative purposes. Note that in all problem instances except #2, only one best result has been produced—that is, we do not see any obvious conflict in the two cost functions. The values in the column labelled "maximum" are used for normalisation purposes, and represent the highest cost values seen in any solution returned by the two algorithms. For both algorithms, the table then displays two statistics: the mean size of the solution lists $L$, and the distances between each of the solutions in $L$ and the "best" solutions. This latter performance measure is based on the metric suggested by Deb et al. (2000), which assesses the performance of a multiobjective algorithm by considering the distance between the costs of each solution in $L$ and the costs of some optimal (or near-optimal) solution to the problem. Because the costs of the global optima are not known for these instances, we choose to use the values given in the "best" column as approximations to this. In our case distances are calculated as follows. First, the costs of all solutions returned by the algorithms, in addition to the costs of the "best" solutions, are normalised to values in $[0, 1]$ by dividing by the maximum cost values specified in the table. Next, for each solution, the (Euclidean) distance between the normalised best costs and all normalised solution costs is calculated. The mean, standard deviation and median of all distances returned for each algorithm are recorded in the table.

The results in Table 7.3 can be split into two cases. The first involves the larger problem instances #3 to #10. Here, we see that the multiobjective approach consistently produces better results than the multi-stage approach, which is reflected in

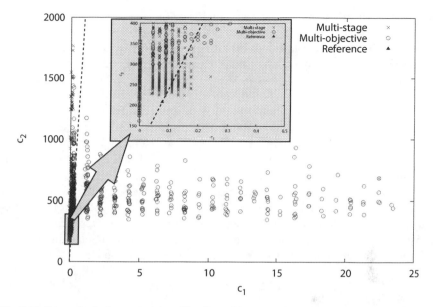

**Fig. 7.14** Costs of solutions encountered by the multi-stage and multiobjective approaches in one run with problem instance #5. Solution costs were recorded every 10,000 evaluations. The dotted line (in both the main graph and the projection) represents the reference line used by the multiobjective approach and is drawn from the origin and through the reference point

**Table 7.3** Summary of results achieved in 100 runs of the multi-stage and multiobjective algorithms. An asterisk (*) in the "best" column indicates that the solution with the associated costs was found using the multi-stage approach (otherwise it was found by the multiobjective approach)

| # | Best $(c_1, c_2)$ | Maximum $c_1$ | $c_2$ | Multi-stage $|L|$ | Distance from Best Mean±SD | Med. | Multiobjective $|L|$ | Distance from Best Mean±SD | Med. |
|---|---|---|---|---|---|---|---|---|---|
| 1 | (0, 104)* | 4.21 | 352 | 1.81 | 0.15±0.22 | 0.04 | 2.57 | 0.27±0.10 | 0.27 |
| 2 | (0, 134)* (2.15, 132)* | 11.3 | 394 | 2.81 | 0.06±0.05 | 0.05 | 3.82 | 0.20±0.11 | 0.17 |
| 3 | (0, 102) | 8.24 | 560 | 4.32 | 0.24±0.17 | 0.17 | 3.32 | 0.13±0.10 | 0.12 |
| 4 | (0, 112) | 2.17 | 186 | 2.72 | 0.28±0.09 | 0.26 | 1.00 | 0.14±0.06 | 0.13 |
| 5 | (0, 134) | 3.29 | 294 | 3.61 | 0.34±0.11 | 0.31 | 1.00 | 0.11±0.04 | 0.11 |
| 6 | (0, 148) | 4.26 | 444 | 4.44 | 0.32±0.11 | 0.30 | 1.04 | 0.08±0.05 | 0.08 |
| 7 | (0, 156) | 2.18 | 816 | 4.51 | 0.19±0.11 | 0.16 | 1.04 | 0.05±0.04 | 0.04 |
| 8 | (0, 186) | 3.27 | 786 | 4.71 | 0.26±0.12 | 0.22 | 1.53 | 0.08±0.06 | 0.06 |
| 9 | (0, 204) | 1.26 | 940 | 4.32 | 0.21±0.10 | 0.19 | 1.08 | 0.05±0.06 | 0.04 |
| 10 | (0, 234) | 2.29 | 1398 | 4.63 | 0.20±0.10 | 0.16 | 1.30 | 0.09±0.05 | 0.07 |

the lower means, medians, and deviations in the Distance from Best column. Note that the best results for these instances have also come from the multiobjective approach. We also see that the multi-stage approach has produced larger solution lists for these problem instances, which could be useful if a user wanted to be presented with a choice of solutions, though the solutions in these lists are of lower quality in general. Also note that the differences between the mean and median values here reveal that the distributions of distances with the multi-stage approach feature larger amounts of positive skew, reflecting the fact that this method produces solutions of very low quality on occasion. For the smaller problem instances #1 and #2 we see that these patterns are more or less reversed from those of the multi-stage approach, producing solutions with costs that are consistently closer (or equal to) to the best known costs. One feature to note in this case is the relatively large difference between the mean and median with the multi-stage approach for instance #1, where the we saw about 60% of produced solutions being very close to the best, and the remainder having much larger distances. Finally, note that for all problem instances, the nonparametric Mann-Whitney test indicates that the distances of each algorithm are significantly different with significance level $\leq 0.01\%$.

In summary, the results in Table 7.3 suggest that the strategy used by the multi-stage algorithm of employing many rounds of short intensive searches seems more fitting for smaller, less constrained instances, but for larger instances, including the real-world problem instance, better solutions are achieved by using the multiobjective approach where longer, less intensive searches are performed.

## 7.7 Chapter Summary and Discussion

In this chapter we have shown that round-robin schedules can be successfully constructed using graph colouring principles, often in the presence of many additional hard constraints. In Section 7.6 we exploited this link with graph colouring by proposing two algorithms for a real-world sports scheduling problem. In the case of the real-world problem instance (#5), we found that more than 98% of all solutions generated by our multiobjective approach dominated the solution that was manually produced by the WRU's league administrators. On the other hand, for the multi-stage approach this figure was just 0.02%. We should, however, interpret these statistics with care, firstly because the manually produced solution was actually for a slightly different problem (the exact specifications of which we were unable to obtain from the league organisers), and secondly because our specific cost functions were not previously used by the league organisers for evaluating their solutions.

One further neighbourhood operator that might be used with these problems (and indeed any graph colouring problem) is an extension to the Kempe chain interchange operator known as the *s-chain* interchange operator. Let $S = \{S_1, \ldots, S_k\}$ be a feasible solution and let $v$ be an arbitrary vertex in $S$ coloured with colour $j_1$. Furthermore, let $j_2, \ldots, j_s$ be a sequence of distinct colours taken from the set $\{1, \ldots, k\} - \{j_1\}$. An $s$-chain is constructed by first identifying all vertices adjacent to $v$ that are coloured with colour $j_2$. From these, adjacent vertices coloured with $j_3$ are then identified, and from these adjacent vertices with colour $j_4$, and so on. When considering vertices with colour $j_s$, adjacent vertices with colour $j_1$ are sought.

As an example, using the graph from Figure 6.3, an $s$-chain using $s = 3$, $v = v_2$ and colours $j_1 = 2$, $j_2 = 4$, and $j_3 = 1$ can be seen to contain the vertices $\{v_2, v_3, v_4, v_5, v_7, v_8\}$. Through similar reasoning to that of Kempe chains (Theorem 3.1), it is simple to show that we can take the vertices of an $s$-chain and interchange their colours using the mapping $j_1 \leftarrow j_2, j_2 \leftarrow j_3, \ldots, j_{s-1} \leftarrow j_s, j_s \leftarrow j_1$ such that feasibility of the solution is maintained. Note also that $s$-chains are equivalent to Kempe chains when $s = 2$. In our experiments with round-robin schedules, we also tested the effects of the $s$-chain interchange operator for $s \geq 3$; however, because of the relatively high levels of connectivity between different colours in these graphs, we observed that over 99% of moves contained the maximum $s(t/2)$ vertices. In other words, almost all moves simply resulted in moves that are also achievable through combinations of $N_4$. $s$-chains may show more promise in other applications however, particularly those involving sparser graphs.

# Chapter 8
# Designing University Timetables

In this chapter, our final case study looks at how graph colouring concepts can be used in the construction of high-quality timetables for universities and other types of educational establishments. As we will see, this sort of problem can contain a whole host of different, and often idiosyncratic, constraints which will often make the problem very difficult to tackle. That said, most timetabling problems contain an underlying graph colouring problem, allowing us to use many of the concepts developed in previous chapters.

The first section of this chapter will look at university timetabling from a broad perspective, discussing amongst other things the various constraints that might be imposed on the problem. Section 8.2 onwards will then conduct a detailed analysis of a well-known timetabling formulation that has been the subject of various articles in the literature. As we will see, powerful algorithms derived from graph colouring principles can be developed for this problem, though careful modifications also need to be made in order to allow the methods to cope with the various other constraints that this problem involves.

## 8.1 Problem Background

In the context of higher education institutions, a timetable can be thought of as an assignment of events (such as lectures, tutorials, or exams) to a finite number of rooms and timeslots in accordance with a set of constraints, some of which will be mandatory, and others that may be optional. It is suggested by Corne et al. (1995) that timetabling constraints can be categorised into five main classes:

Unary Constraints that involve just one event, such as the constraint "event $a$ must not take place on a Tuesday", or the constraint "event $a$ must occur in timeslot $b$".

© Springer International Publishing Switzerland 2016
R.M.R. Lewis, *A Guide to Graph Colouring*,
DOI 10.1007/978-3-319-25730-3_8

Binary Constraints    that concern *pairs* of events, such as the constraint "event *a* must take place before event *b*", or the *event clash* constraint, which specifies pairs of events that cannot be held at the same time in the timetable.

Capacity Constraints    that are governed by room capacities, etc. For example "All events should be assigned to a room that has a sufficient capacity".

Event Spread Constraints    that concern requirements involving the "spreading-out" or "grouping-together" of events within the timetable in order to ease student/teacher workload, and/or to agree with a university's timetabling policy.

Agent Constraints    that are imposed in order to promote the requirements and/or preferences of the people who will use the timetables, such as the constraint "lecturer *a* likes to teach event *b* on Mondays", or "lecturer *c* must have *d* free mornings per week".

Like many problems in operational research, a convention in automated timetabling is to group constraints into two classes: *hard* constraints and *soft* constraints. Hard constraints have a higher priority than soft, and their satisfaction is usually mandatory. Indeed, timetables will normally only be considered "feasible" if *all* of the imposed hard constraints have been satisfied. Soft constraints, meanwhile, are those that we want to obey *if possible*, and more often than not they will describe the criteria for a timetable to be "good" with regard to the timetabling policies of the university concerned, as well as the experiences of the people who will have to use it.

According to McCollum et al. (2010), the problem of constructing university timetables can be divided into two categories: exam timetabling problems and course timetabling problems. It is also suggested that course timetabling problems can be further divided into two subcategories: "post enrolment-based course timetabling", where the constraints of the problem are specified by student enrolment data, and "curriculum-based course timetabling", where constraints are based on curricula specified by the university. Müller and Rudova (2012) have also shown that these subcategories are closely related, demonstrating how instances of the latter can be transformed into those of the former in many cases.

We have seen in Sections 1.1.2 and 4.2.2 that a fundamental constraint in university timetabling is the "event-clash" binary constraint. This specifies that if a person (or some other resource of which there is only one) is required to be present in a pair of events, then these must not be assigned to the same timeslot, as such an assignment will result in this person/resource having to be in two places at once. This constraint can be found in almost all university timetabling problems and allows us to draw parallels with the graph colouring problem by considering the events as vertices, clashing events as adjacent vertices, and the timeslots as colours. Beyond this near-universal constraint however, timetabling problem formulations will usually vary widely from place to place due to the fact that different universities will usually have their own individual needs and timetabling policies (and therefore set of constraints) that they need to satisfy (see Figure 8.1). That said, it is still the case that nearly all timetabling problems do feature the underlying graph colouring problem in some form or another in their definitions, and many timetabling algorithms do use various pieces of heuristic information extracted from the graph colouring problem as a driving force in their searches for solutions.

**Fig. 8.1** When constructing a timetable, meeting the needs of all concerned may not always be possible

Despite this wide range of constraints, it is known that university timetabling problems are NP-complete in almost all variants. Cooper and Kingston (1996), for example, have proved that NP-completeness exists for a number of different problem interpretations that can arise in practice. This they achieve by providing polynomial transformations from various NP-complete problems, including bin packing, three-dimensional matching as well as graph colouring itself.

## 8.1.1 Designing and Comparing Algorithms

The field of university timetabling has seen many solution approaches proposed over the years, including methods based on constructive heuristics, mathematical programming, branch and bound, and metaheuristics. (See, for example, the surveys of Carter et al. (1996), Burke et al. (1996), Schaerf (1999), and Lewis (2008).) The latter survey has suggested that metaheuristic approaches for university timetabling can be loosely classified into three categories as follows:

One-stage Optimisation Algorithms.    Here, the satisfaction of the hard constraints and soft constraints is attempted simultaneously, usually using a single objective function in which violations of hard constraints are penalised more heavily than violations of soft constraints. If desired, these weights could be altered during a run.

Two-stage Optimisation Algorithms.    In this case, the hard constraints are first sat-
    isfied to form a feasible solution. Attempts are then made to eliminate violations
    of the soft constraints by navigating the space of feasible solutions. (Note that
    similar schemes have been used in the case studies from Chapters 6 and 7.)
Algorithms That Allow Relaxations.    Here, violations of the hard constraints are
    disallowed from the outset by relaxing some features of the problem. Attempts
    are then made to try and satisfy the soft constraints, whilst also giving considera-
    tion to the task of eliminating these relaxations. These relaxations could include
    allowing certain events to be left out of the timetable, or using additional times-
    lots or rooms.

The wide variety of constraints, coupled with the fact that each higher education institution will usually have their own specific timetabling needs and protocols, means that timetabling problem formulations have always tended to vary quite
widely in the literature. While making the problem area very rich, one drawback
has been the lack of opportunity for accurate comparison of algorithms over the
years. Since the early 2000s, this situation has been mitigated to a certain extent
with the organisation of a series of timetabling competitions and the release of publicly available problem instances.[1] In 2007, for example, the Second International
Timetabling Competition (ITC2007) was organised by a group of timetabling researchers from different European Universities, which considered the three types
of timetabling problems mentioned above: exam timetabling, post enrolment-based
course timetabling, and curriculum-based timetabling. The competition operated by
releasing problem instances into the public domain, with entrants then designing
algorithms to try and solve these. Entrants' algorithms were then compared under
strict time limits according to specific evaluation criteria.

## 8.1.2  Chapter Outline

In this chapter we will examine the post enrolment-based course timetabling problem used for ITC2007. This formulation models the real-world situation where students are given a choice of lectures that they wish to attend, with the timetable
then being constructed according to these choices. The next section contains a formal definition of this problem, with Section 8.3 then containing a short review of the
most noteworthy algorithms. We then go on to describe our method and its operators
in Sections 8.4 and 8.5. The final results of our algorithm are given in Section 8.6,
with a discussion and conclusions then being presented in Section 8.7.

---

[1] The official websites of these competitions can be found at www.cs.qub.ac.uk/itc2007/ and
www.utwente.nl/ctit/hstt/itc2011.

## 8.2 Problem Definition and Preprocessing

As mentioned, the post enrolment-based course timetabling problem was introduced for use in the Second International Timetabling Competition, run in 2007. The problem involves seven types of hard constraint whose satisfaction is mandatory, and three soft constraints, whose satisfaction is desirable, but not essential. The problem involves assigning a set of events to 45 timeslots (five days, with nine timeslots per day) according to these constraints.

The hard constraints for the problem are as follows. First, for each event there is a set of students who are enrolled to attend. Events should be assigned to timeslots such that no student is required to attend more than one event in any one timeslot. Next, each event also requires a set of room features (e.g., a certain number of seats, specialist teaching equipment, etc.), which will only be provided by certain rooms. Consequently, each event needs to be assigned to a suitable room that exhibits the room features that it requires. The double booking of rooms is also disallowed. Hard constraints are also imposed stating that some events cannot be taught in certain timeslots. Finally, precedence constraints—stating that some events need to be scheduled before or after others—are also stipulated.

More formally, a problem instance comprises a set of events $e = \{e_1, \ldots, e_n\}$, a set of timeslots $t = \{t_1, \ldots, t_k\}$ (where $k = 45$), a set of students $s = \{s_1, \ldots, s_{|s|}\}$, a set of rooms $r = \{r_1, \ldots, r_{|r|}\}$, and a set of room features $f = \{f_1, \ldots, f_{|f|}\}$. Each room $r_i \in r$ is also allocated a capacity $c(r_i)$ reflecting the number of seats it contains.

The relationships between the above sets are defined by five problem matrices: an *attends* matrix $\mathbf{P}^{(1)}_{|s| \times n}$ where

$$P^{(1)}_{ij} = \begin{cases} 1 & \text{if student } s_i \text{ is due to attend event } e_j \\ 0 & \text{otherwise,} \end{cases}$$

a *room features* matrix $\mathbf{P}^{(2)}_{|r| \times |f|}$ where

$$P^{(2)}_{ij} = \begin{cases} 1 & \text{if room } r_i \text{ has feature } f_j \\ 0 & \text{otherwise,} \end{cases}$$

an *event features* matrix $\mathbf{P}^{(3)}_{n \times |f|}$ where

$$P^{(3)}_{ij} = \begin{cases} 1 & \text{if event } e_i \text{ requires feature } f_j \\ 0 & \text{otherwise,} \end{cases}$$

an *event availability* matrix $\mathbf{P}^{(4)}_{n \times k}$ in which

$$P_{ij}^{(4)} = \begin{cases} 1 & \text{if event } e_i \text{ can be assigned to timeslot } t_j \\ 0 & \text{otherwise,} \end{cases}$$

and finally a *precedence* matrix $\mathbf{P}_{n \times n}^{(5)}$ where

$$P_{ij}^{(5)} = \begin{cases} 1 & \text{if event } e_i \text{ should be assigned to an earlier timeslot than event } e_j \\ -1 & \text{if event } e_i \text{ should be assigned to a later timeslot than event } e_j \\ 0 & \text{otherwise.} \end{cases}$$

For the precedence matrix above, note that two conditions are necessary for the relationships to be consistent: (a) $P_{ij}^{(5)} = 1$ if and only if $P_{ji}^{(5)} = -1$, and (b) $P_{ij}^{(5)} = 0$ if and only if $P_{ji}^{(5)} = 0$. We can also observe the transitivity of this relationship:

$$\left( \exists e_i, e_j, e_l \in e : \left( P_{ij}^{(5)} = 1 \wedge P_{jl}^{(5)} = 1 \right) \right) \Rightarrow P_{il}^{(5)} = 1 \qquad (8.1)$$

In some of the competition problem instances this transitivity is not fully expressed; however, observing it enables further 1's and $-1$'s to be added to $\mathbf{P}^{(5)}$ during pre-processing, allowing the relationships to be more explicitly stated.

Given the above five matrices, we are also able to calculate two further matrices that allow fast detection of hard constraint violations. The first of these is a *room suitability* matrix $\mathbf{R}_{n \times |r|}$ defined as:

$$R_{ij} = \begin{cases} 1 & \text{if } \left( \sum_{l=1}^{|s|} P_{li}^{(1)} \leq c(r_j) \right) \wedge \left( \nexists f_l \in f : \left( P_{il}^{(3)} = 1 \wedge P_{jl}^{(2)} = 0 \right) \right) \\ 0 & \text{otherwise.} \end{cases} \qquad (8.2)$$

The second is then a *conflicts* matrix $\mathbf{C}_{n \times n}$, defined:

$$C_{ij} = \begin{cases} 1 & \text{if } \left( \exists s_l \in s : \left( P_{li}^{(1)} = 1 \wedge P_{lj}^{(1)} = 1 \right) \right) \\ & \vee \left( \left( \exists r_l \in r : \left( R_{il} = 1 \wedge R_{jl} = 1 \right) \right) \wedge \left( \sum_{l=1}^{|r|} R_{il} = 1 \right) \wedge \left( \sum_{l=1}^{|r|} R_{jl} = 1 \right) \right) \\ & \vee \left( P_{ij}^{(5)} \neq 0 \right) \\ & \vee \left( \nexists t_l \in t : \left( P_{il}^{(4)} = 1 \wedge P_{jl}^{(4)} = 1 \right) \right) \\ 0 & \text{otherwise.} \end{cases}$$

$$(8.3)$$

The matrix $\mathbf{R}$ therefore specifies the rooms that are suitable for each event (that is, rooms that are large enough for all attending students and that have all the required features). The $\mathbf{C}$ matrix, meanwhile, is a symmetrical matrix ($C_{ij} = C_{ji}$) that specifies pairs of events that cannot be assigned to the same timeslot (i.e., those that conflict). According to Equation (8.3) this will be the case if two events $e_i$ and $e_j$ share a common student, require the same individual room, are subject to a precedence relation, or have mutually exclusive subsets of timeslots for which they are available.

Note that the matrix $\mathbf{C}$ is analogous to the adjacency matrix of a graph $G = (V, E)$ with $n$ vertices, highlighting the similarities between this timetabling problem and the graph colouring problem. However, unlike the graph colouring problem, in this case the ordering of the timeslots (colour classes) is also an important property of a solution. Consequently, a solution is represented by an ordered set of sets $S = (S_1, \ldots, S_{k=45})$ and is subject to the satisfaction of the following hard constraints.

$$\bigcup_{i=1}^{k} S_i \subseteq e \tag{8.4}$$

$$S_i \cap S_j = \emptyset \quad (1 \le i \ne j \le k) \tag{8.5}$$

$$\forall e_j, e_l \in S_i, C_{jl} = 0 \quad (1 \le i \le k) \tag{8.6}$$

$$\forall e_j \in S_i, P_{ji}^{(4)} = 1 \quad (1 \le i \le k) \tag{8.7}$$

$$\forall e_j \in S_i, e_l \in S_{q<i}, P_{jl}^{(5)} \ne 1 \quad (1 \le i \le k) \tag{8.8}$$

$$\forall e_j \in S_i, e_l \in S_{q>i}, P_{jl}^{(5)} \ne -1 \quad (1 \le i \le k) \tag{8.9}$$

$$S_i \in \mathcal{M} \quad (1 \le i \le k) \tag{8.10}$$

Constraints (8.4) and (8.5) state that $S$ should partition the event set $e$ (or a subset of $e$) into an ordered set of sets, labelled $S_1, \ldots, S_k$. Each set $S_i \in S$ contains the events that are assigned to timeslot $t_i$ in the timetable. Constraint (8.6) stipulates that no pair of conflicting events should be assigned to the same set $S_i \in S$, while Constraint (8.7) states that each event should be assigned to a set $S_i \in S$ whose corresponding timeslot $t_i$ is deemed available according to matrix $\mathbf{P}^{(4)}$. Constraints (8.8) and (8.9) then impose the precedence requirements of the problem.

Finally, Constraint (8.10) is concerned with ensuring that the events assigned to a set $S_i \in S$ can each be assigned to a suitable room from the room set $r$. To achieve this it is necessary to solve a maximum bipartite matching problem. Specifically, let $G = (S_i, r, E)$ be a bipartite graph with vertex sets $S_i$ and $r$, and an edge set: $E = \{\{e_j \in S_i, r_l \in r\} : R_{jl} = 1\}$. Given $G$, the set $S_i$ is a member of $\mathcal{M}$ if and only if there exists a maximum bipartite matching of $G$ of size $2|S_i|$, in which case the room constraints for this timeslot will have been satisfied.

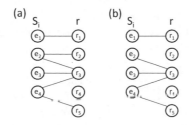

**Fig. 8.2** Example bipartite graphs with and without a maximum bipartite matching of size $2|S_i|$. In Part (a), events $e_1, e_2, e_3$, and $e_4$ can be assigned to rooms $r_1, r_2, r_3$, and $r_5$ respectively

Figure 8.2 shows two examples of these ideas using $|S_i| = 4$ events and $|r| = 5$ rooms. In Figure 8.2(a) a matching exists—for example, event $e_1$ can be assigned to room $r_1$, $e_2$ to $r_2$, $e_3$ to $r_3$, and $e_4$ to $r_5$. On the other hand, a matching does not exist in Figure 8.2(b), meaning $S_i \notin \mathcal{M}$ in this case. In practice, such matching problems can be solved via various polynomially bounded algorithms such as the augmenting path, Hungarian, or auction algorithms (Bertsekas, 1992).

In the competition's interpretation of this problem, a solution $\mathcal{S}$ is considered *valid* if and only if all constraints (8.4)–(8.10) are satisfied. The quality of a valid solution is then gauged by a *distance to feasibility* (DTF) measure, calculated as the sum of the sizes of all events not contained in the solution:

$$\text{DTF} = \sum_{e_i \in S'} \sum_{j=1}^{|s|} P_{ij}^{(1)} \tag{8.11}$$

where $S' = e - (\bigcup_{i=1}^{k} S_i)$. If the solution $\mathcal{S}$ is valid and has a DTF of zero (implying $\bigcup_{i=1}^{k} S_i = e$ and $S' = \emptyset$) then it is considered *feasible* since all of the events have been feasibly timetabled. The set of feasible solutions is thus a subset of the set of valid solutions.

## 8.2.1 Soft Constraints

As mentioned, in addition to finding a solution that obeys all of the hard constraints, three soft constraints are also considered in this problem.

SC1    Students should not be required to attend an event in the last timeslot of each day (i.e., timeslots 9, 18, 27, 36, or 45);

SC2    Students should not have to attend events in three or more successive times-lots occurring in the same day; and,

SC3    Students should not be required to attend just one event in a day.

The extent to which these constraints are violated is measured by a Soft Constraints Cost (SCC), which is worked out in the following way. For SC1, if a student attends an event assigned to an end-of-day timeslot, this is counted as one penalty point. Naturally, if $x$ students attend this class, this counts as $x$ penalty points. For SC2 meanwhile, if a student attends three events in a row we count this as one penalty point. If a student has four events in a row we count this as two, and so on. Note that students assigned to events occurring in consecutive timeslots over two separate days are not counted as violations. Finally, each time we encounter a student with a single event on a day, we count this as one penalty point (two for two days with single events, etc.). The SCC is simply the total of these three values.

More formally, the SCC can be calculated using two matrices: $\mathbf{X}_{|s| \times k}$, which tells us the timeslots for which each student is attending an event, and $\mathbf{Y}_{|s| \times 5}$, which specifies whether or not a student is required to attend just one event in each of the five days.

$$X_{ij} = \begin{cases} 1 & \text{if } \exists e_l \in S_j : P_{il}^{(1)} = 1 \\ 0 & \text{otherwise,} \end{cases} \qquad (8.12)$$

$$Y_{ij} = \begin{cases} 1 & \text{if } \sum_{l=1}^{9} X_{i,9(j-1)+l} = 1 \\ 0 & \text{otherwise.} \end{cases} \qquad (8.13)$$

Using these matrices, the SCC is calculated as follows:

$$\text{SCC} = \sum_{i=1}^{|s|} \sum_{j=1}^{5} \left( (X_{i,9j}) + \left( \sum_{l=1}^{7} \prod_{q=0}^{2} X_{i,9(j-1)+l+q} \right) + (Y_{i,j}) \right). \qquad (8.14)$$

Here the three terms summed in the outer parentheses of Equation 8.14 define the number of violations of SC1, SC2, and SC3 (respectively) for each student on each day of the timetable.

## 8.2.2 Problem Complexity

Having defined the post enrolment-based course timetabling problem, we are now in a position to state its complexity.

**Theorem 8.1** *The post enrolment-based course timetabling problem is NP-hard.*

*Proof.* Let $\mathbf{C}_{n \times n}$ be our symmetric conflicts matrix as defined above, filled arbitrarily. In addition, let the following conditions hold:

$$|r| \geq n \qquad (8.15)$$
$$R_{lj} = 1 \ \forall e_i \in e, r_j \in r \qquad (8.16)$$
$$P_{ij}^{(4)} = 1 \ \forall e_i \in e, t_j \in t \qquad (8.17)$$
$$P_{ij}^{(5)} = 0 \ \forall e_i, e_j \in e \qquad (8.18)$$

Here, there is an excess number of rooms which are suitable for all events (8.15-8.16), there are no event availability constraints (8.17), and no precedence constraints (8.18). In this special case we are therefore only concerned with satisfying Constraints (8.4)–(8.6) while minimising the DTF. Determining the existence of a feasible solution using $k$ timeslots is therefore equivalent to the NP-complete graph $k$-colouring problem. □

From a different perspective, Camhazard et al. (2012) have also shown that, in the absence of all hard constraints (8.6)–(8.10) and soft constraints SC1 and SC2, the problem of satisfying SC3 (i.e., minimising the number of occurrences of students sitting a single event in a day) is equivalent to the NP-hard set covering problem.

## 8.2.3 Evaluation and Benchmarking

From the above descriptions, we see that a timetable's quality is described by two values: the distance to feasibility (DTF) and the soft constraint cost (SCC). According to the competition criteria, when comparing solutions the one with the lowest DTF is deemed the best timetable, reflecting the increased importance of the hard constraints over the soft constraints. However, when two or more solutions' DTFs are equal, the winner is deemed the solution among these that has the lowest SCC.

Currently there are 24 problem instances available for this problem, all of which are known to feature at least one perfect solution (that is, a solution with DTF = 0 and SCC = 0). For comparative purposes, a benchmark timing program is also available on the competition website that allocates a strict time limit for each machine that it is executed on (based on its hardware and operating system). This allows researchers to use approximately the same amount of computational effort when testing their implementations, allowing more accurate comparisons.

## 8.3 Previous Approaches to This Problem

One of the first studies into the post enrolment-based timetabling problem (in this form) was carried out by Rossi-Doria et al. (2002), who used it as a test problem for comparing five different metaheuristics, namely evolutionary algorithms, simulated annealing, iterated local search, ant colony optimisation, and tabu search. Two interesting observations were offered in their work:

- "The performance of a metaheuristic with respect to satisfying hard constraints and soft constraints may be different."
- "Our results suggest that a hybrid algorithm consisting of at least two phases, one for taking care of feasibility, the other taking care of minimising the number of soft constraint violations [without reviolating any of the hard constraints in the process], is a promising direction." (Rossi-Doria et al., 2002)

These conclusions have proven to be quite salient in the past decade, with a number of successful algorithms following this suggested two-stage methodology. This includes the winning entry of ITC2007 itself, due to Cambazard et al. (2008), which uses tabu search together with an intensification procedure to achieve feasibility, with simulated annealing then being used to satisfy the soft constraints.

Since the running of these competitions, a number of papers have been published that have equalled or improved upon the results of the 2007 competition. Cambazard et al. themselves have shown how the results of their two-stage competition entry can be improved by relaxing Constraint (8.10) such that a timeslot $t_i$ is considered feasible whenever $|S_i| < |r|$ (Cambazard et al., 2012). The rationale for this relaxation is that it will "increase the solution density of the underlying search space", though a repair operator is also needed to make sure that the timeslots satisfy Constraint (8.10) at the end of execution. Cambazard et al. (2012) have also examined constraint programming-based approaches and a large neighbourhood search (LNS) scheme, and find that their best results can be found when using simulated annealing together with the LNS operator for reinvigorating the search from time to time.

Other successful algorithms for this problem have followed the one-stage optimisation scheme by attempting to reduce violations of hard and soft constraints simultaneously. Ceschia et al. (2012), for example, treat this problem as a single objective optimisation problem in which the space of valid *and invalid* solutions is explored. Specifically, they allow violations of Constraints (8.6), (8.8), and (8.9) within a solution, and use the number of students affected by such violations, together with the DTF, to form an infeasibility measure. This is then multiplied by a weighting coefficient $w$ and added to the SCC to form the objective function. Simulated annealing is then used to optimise this objective function and, surprisingly, after extensive parameter tuning $w = 1$ is found to provide their best results. Nothegger et al. (2012) have also attempted to optimise the DTF and SCC simultaneously, making use of ant colony optimisation to explore the space of valid solutions. Here the DTF and SCC measures are used to update the algorithm's pheromone matrices so that favourable assignments of events to rooms and timeslots will occur with higher probability in later iterations of the algorithm. Nothegger et al. also show that the results of their algorithm can be improved by adding a local search-based improvement method and by parallelising the algorithm. Jat and Yang (2011) have also used a weighted sum objective function in their hybrid evolutionary algorithm/tabu search approach, though their results do not appear as strong as those of the previous two papers. Similarly, van den Broek and Hurkens (2012) have also used a weighted sum objective function in their deterministic algorithm based on column generation techniques.

From the above studies, it is clear that the density and connectivity of the underlying solution space is an important issue in the performance of a neighbourhood search algorithm with this problem. In particular, if connectivity is low then movements in the solution space will be more restricted, perhaps making improvements in the objective function more difficult to achieve. From the research discussed above it is noticeable that some of the best approaches for this problem have attempted to alleviate this problem by relaxing some of the hard constraints and/or by allowing events to be kept out of the timetable. However, such methods also require mechanisms for coping with these relaxations, such as repair operators (which may ultimately require large alterations to be made to a solution), or by introducing terms into the objective function (which will require appropriate weighting coefficients to be determined, perhaps via tuning). On the other hand, a two-stage approach of the type discussed by Rossi-Doria et al. (2002) will have no need for these features, though because feasibility must be maintained when the SCC is being optimised, the underlying solution space may be more sparsely connected, perhaps making good levels of optimisation more difficult to achieve. We will focus on the issue of connectivity in Section 8.5 onwards.

## 8.4 Algorithm Description: Stage One

Before looking at the task of eliminating soft constraint violations it is first necessary to produce a valid solution that minimises the DTF measure (Equation (8.11)). Previous strategies for this task have typically involved inserting all events into the timetable, and then rearranging them in order to remove violations of the hard constraints (Cambazard et al., 2008, 2012; Ceschia et al., 2012; Chiarandini et al., 2006). In contrast, we suggest using a method by which events are permitted to remain outside of the timetable, meaning that spaces within the timetable are not "blocked" by events causing hard constraint violations. To do this, we exploit the similarity between this problem and the graph colouring problem by using an adaptation of the PARTIALCOL algorithm from Section 4.1.2.

To begin, an initial solution is constructed by taking events one by one and assigning them to timeslots such that none of the hard constraints are violated. Events that cannot be assigned without breaking a hard constraint are kept to one side and are dealt with at the end of this process. To try and maximise the number of events inserted into the timetable, a set of high performance heuristics originally proposed by Lewis (2012) and based on the DSATUR algorithm (Section 2.3) is used. At each step heuristic rule $h_1$ (Table 8.1) is used to select an event, with ties being broken using $h_2$, and then $h_3$ (if necessary). The selected event is then inserted into the timetable according to rule $h_4$, breaking ties with $h_5$ and further ties with $h_6$.

**Table 8.1** Heuristics used for producing an initial solution in Stage 1. Here, a "valid place" is defined as a room/timeslot pair that an event can be assigned to without violating Constraints (1.1)–(8.10)

| Rule | Description |
|------|-------------|
| $h_1$ | Choose the unplaced event with the smallest number of valid places in the timetable to which it can be assigned. |
| $h_2$ | Choose the unplaced event $e_i$ that conflicts with the most other events (i.e that maximises $\sum_{j=1}^{n} C_{i,j}$). |
| $h_3$ | Choose an event randomly. |
| $h_4$ | Choose the place that is valid for the least number of other unplaced events in $U$. |
| $h_5$ | Choose the valid place in the timeslot containing the fewest events. |
| $h_6$ | Choose a place randomly. |

Completion of this constructive phase results in a valid solution $S$ obeying Constraints (8.4)–(8.10). However, if $S' \neq \emptyset$ (where $S' = e - (\bigcup_{i=1}^{k} S_i)$ is the set of unplaced events), the PARTIALCOL algorithm will need to be invoked.

As with the original algorithm, this method operates using tabu search with the simple cost function $|S'|$. During a run the neighbourhood operator moves events between $S'$ and timeslots in $S$ while maintaining the validity of the solution. Given an event $e_i \in S'$ and timeslot $S_j \in S$, checks are first made to see if $e_i$ can be assigned to $S_j$ without violating Constraints (8.7)–(8.9). If the assignment of $e_i$ violates one of these constraints, the move is rejected; else, all events $e_k$ in $S_j$ that conflict with $e_i$

(according to the conflicts matrix $\mathbf{C}$) are transferred from $S_j$ into $S'$ and an attempt is made to insert event $e_i$ into $S_j$, perhaps using a maximum matching algorithm. If this is not possible (i.e., Constraint (8.10) cannot be satisfied) then all changes are again reset; otherwise the move is considered valid. Upon performing this change, all moves involving the reassignment of event(s) $e_k$ to timeslot $S_j$ are considered tabu for a number of iterations. Here we use the same tabu tenure as that of the TABUCOL algorithm seen in Chapter 4. Hence the tenure is defined as a random variable proportional to the current cost, $\lfloor 0.6 \times |S'| + x \rfloor$, where $x$ is an integer uniformly selected from the set $\{0, 1, \ldots, 9\}$.

Similarly to the original PARTIALCOL algorithm, at each iteration the entire neighbourhood of $(|S'| \times k)$ moves is examined, and the move that is chosen is the one that invokes the largest decrease (or failing that, the smallest increase) in the cost of any valid, non-tabu move. Ties are broken randomly, and tabu moves are also permitted if they are seen to improve on the best solution found so far. From time to time, there may also be no valid non-tabu moves available from a particular solution, in which case a randomly selected event is transferred from $S$ into $S'$, before the process is continued as above.

## 8.4.1 Results

Table 8.2 contains the results of our PARTIALCOL algorithm and compares them to those reported by Cambazard et al. (2012).[2] We report the percentage of runs in which each instance has been solved (i.e., where a DTF of zero has been achieved), and the average time that this took (calculated only from the solved runs).[3] We see that the success rates for the two approaches are similar, with all except one instance being solved in 100% of cases (instance #10 in Cambazard et al.'s case, instance #11 in ours). However, with the exception of instance #11, the time required by PARTIALCOL is considerably less, with an average reduction of 97.4% in CPU time achieved across the 15 remaining instances.

Curiously, when using our PARTIALCOL algorithm with instance #11, most of the runs were solved very quickly. However, in a small number of runs the algorithm seemed to quickly navigate to a point at which a small number of events remained unplaced and where no further improvements could be made, suggesting the search

---

[2] Our algorithm was implemented in C++, and all experiments were conducted on 3.0 GHz Windows 7 PCs with 3.87 GB RAM. The competition benchmarking program allocated 247 s on this equipment. The source code is available at www.rhydlewis.eu/resources/ttCodeResults.zip

[3] For comparative purposes, the computation times stated by Cambazard et al. (2012) have been altered in Table 8.2 to reflect the increased speed of our equipment. Specifically, according to Cambazard et al. the competition benchmark program allocated them 324 s per run. Consequently, their original run times have been reduced by 23.8%. We should note, however, that when comparing algorithms in this way, discrepancies in results and times can also occur due to differences in the hardware, operating system, programming language, and compiler options that are used. Our use of the competition benchmark program attempts to reduce discrepancies caused by the first two factors, but cannot correct for differences arising due to the latter two.

**Table 8.2** Comparison of results from the LS-colouring method of Cambazard et al. (2012), and our PARTIALCOL and Improved PARTIALCOL algorithms (all figures taken from 100 runs per instance)

| | Instance # | 1 | 2 | 3 | 4 | 5 | 6 | 7 | 8 | 9 | 10 | 11 | 12 |
|---|---|---|---|---|---|---|---|---|---|---|---|---|---|
| Cambazard et al. (2012) | % Solved | 100 | 100 | 100 | 100 | 100 | 100 | 100 | 100 | 100 | 98 | 100 | 100 |
| | Avg. time (s) | 11.60 | 37.10 | 0.37 | 0.43 | 3.58 | 4.32 | 1.84 | 1.11 | 51.73 | 170.24 | 0.40 | 0.64 |
| PARTIALCOL | % Solved | 100 | 100 | 100 | 100 | 100 | 100 | 100 | 100 | 100 | 100 | 98 | 100 |
| | Avg. time (s) | 0.25 | 0.79 | 0.02 | 0.04 | 0.05 | 0.07 | 0.02 | 0.01 | 0.71 | 1.80 | 1.88 | 0.04 |
| Improved PARTIALCOL | % Solved | 100 | 100 | 100 | 100 | 100 | 100 | 100 | 100 | 100 | 100 | 100 | 100 |
| | Avg. time (s) | 0.25 | 0.79 | 0.02 | 0.02 | 0.06 | 0.08 | 0.03 | 0.01 | 0.68 | 2.03 | 0.03 | 0.04 |

| | Instance # | 13 | 14 | 15 | 16 | 17 | 18 | 19 | 20 | 21 | 22 | 23 | 24 |
|---|---|---|---|---|---|---|---|---|---|---|---|---|---|
| Cambazard et al. (2012) | % Solved | 100 | 100 | 100 | 100 | - | - | - | - | - | - | - | - |
| | Avg. time (s) | 8.86 | 7.97 | 0.80 | 0.55 | - | - | - | - | - | - | - | - |
| PARTIALCOL | % Solved | 100 | 100 | 100 | 100 | 100 | 100 | 100 | 100 | 100 | 100 | 100 | 100 |
| | Avg. time (s) | 0.08 | 0.11 | 0.01 | 0.01 | 0.00 | 0.02 | 0.74 | 0.01 | 0.07 | 3.77 | 1.33 | 0.17 |
| Improved PARTIALCOL | % Solved | 100 | 100 | 100 | 100 | 100 | 100 | 100 | 100 | 100 | 100 | 100 | 100 |
| | Avg. time (s) | 0.08 | 0.11 | 0.01 | 0.01 | 0.00 | 0.02 | 0.71 | 0.01 | 0.08 | 3.80 | 1.10 | 0.18 |

was caught in a conspicuous valley in the cost landscape. To remedy this situation we therefore added a diversification mechanism to the method which attempts to break out of such regions. We call this our improved PARTIALCOL algorithm and its results are also given in Table 8.2.

In the improved PARTIALCOL method, our diversification mechanism is used for making relatively large changes to the incumbent solution, allowing new regions of the solution space to be explored. It is called when the best solution found so far has not been improved for a set number of iterations. The mechanism operates by first randomly selecting a percentage of events in $S$ and transferring them to the set of unplaced events $S'$. Next, alterations are made to $S$ by performing a random walk using neighbourhood operator $N_5$ (to be described in Section 8.5). Finally, the tabu list is reset so that all potential moves are deemed non-tabu, before PARTIAL-COL continues to execute as before. For the results in Table 8.2, the diversification mechanism was called after 5,000 nonimproving iterations, and extracted 10% of all events in $S$. A random walk of 100 neighbourhood moves was then performed, giving a $> 95\%$ chance of all timeslots being altered by the neighbourhood operator (a number of other parameters were also tried here, though few differences in performance were observed). We see that the improved PARTIALCOL method has achieved feasibility in all runs in the sample, with the average time reduction remaining at 97.4% compared to the method of Cambazard et al. (2012).

## 8.5  Algorithm Description: Stage Two

### 8.5.1  SA Cooling Scheme

In the second stage of this algorithm we use simulated annealing (SA) to explore the space of valid/feasible solutions, and attempt to minimise the number of soft constraint violations measured by the SCC (Equation (8.14)). This metaheuristic is applied in a similar manner to that described in Chapter 3: starting at an initial temperature $T_0$, during execution the temperature variable is slowly reduced according to an update rule $T_{i+1} = \alpha T_i$, where the cooling rate $\alpha \in [0, 1)$. At each temperature $T_i$, a Markov chain is generated by performing $n^2$ applications of the neighbourhood operator. Moves that are seen to violate a hard constraint are immediately rejected. Moves that preserve feasibility but that increase the cost of the solution are accepted with probability $\exp(-|\delta|/T_i)$ (where $\delta$ is the change in cost), while moves that reduce or maintain the cost are always accepted. The initial temperature $T_0$ is calculated automatically by performing a small sample of neighbourhood moves and using the standard deviation of the cost over these moves (van Laarhoven and Aarts, 1987).

Because this algorithm is intended to operate according to a time limit, a value for $\alpha$ is determined automatically so that the temperature is reduced as slowly as possible between $T_0$ and some end temperature $T_{\text{end}}$. This is achieved by allowing $\alpha$ to be modified during a run according to the length of time that each Markov chain takes to generate. Specifically, let $\mu^*$ denote the estimated number of Markov chains that will be completed in the remainder of the run, calculated by dividing the amount of remaining run time by the length of time the most recent Markov chain (operating at temperature $T_i$) took to generate. On completion of the $i$th Markov chain, a modified cooling rate can thus be calculated as:

$$\alpha_{i+1} = (T_{\text{end}}/T_i)^{1/\mu^*} \tag{8.19}$$

The upshot is that the cooling rate will be altered slightly during a run, allowing the user-specified end temperature $T_{\text{end}}$ to be reached at the time limit. Suitable values for $T_{\text{end}}$, the only parameter required for this phase, are examined in Section 8.6.

### 8.5.2  Neighbourhood Operators

We now define a number of different neighbourhood operators that can be used in conjunction with the SA algorithm for this problem. Let $N(S)$ be the set of candidate solutions in the neighbourhood of the incumbent solution $S$. Also, let $\mathbb{S}$ be the set of all valid solutions (i.e., $S \in \mathbb{S}$ if and only if Constraints (8.4)–(8.10) are satisfied). The relationship between the solution space and neighbourhood operator can now be defined by a graph $G = (\mathbb{S}, E)$ with vertex set $\mathbb{S}$ and edge set $E = \{\{S \in \mathbb{S}, S' \in \mathbb{S}\}$ :

$\mathcal{S}' \in N(\mathcal{S})\}$. Note that all of the following neighbourhood operators are reversible, meaning that $\mathcal{S}' \in N(\mathcal{S})$ if and only if $\mathcal{S} \in N(\mathcal{S}')$. Hence edges are expressed as unordered pairs. The various neighbourhood operators are now defined.

$N_1$:    The first neighbourhood operator is based on those used by Lewis (2012) and Nothegger et al. (2012). Consider a valid solution $\mathcal{S}$ represented as a matrix $\mathbf{Z}_{|r| \times k}$ in which rows represent rooms and columns represent timeslots. Each element of $\mathbf{Z}$ can be blank or can be occupied by exactly one event. If $Z_{ij}$ is blank, then room $r_i$ is *vacant* in timeslot $t_j$. If $Z_{ij} = e_l$, then event $e_l$ is assigned to room $r_i$ and timeslot $t_j$. $N_1$ operates by first randomly selecting an element $Z_{i_1 j_1}$ containing an arbitrary event $e_l$. A second element $Z_{i_2 j_2}$ is then randomly selected in a different timeslot ($j_1 \neq j_2$). If $Z_{i_2 j_2}$ is blank, the operator attempts to transfer $e_l$ from timeslot $j_1$ into any vacant room in timeslot $j_2$. If $Z_{i_2 j_2} = e_q$, then a swap is attempted in which $e_l$ is moved into any vacant room in timeslot $j_2$, and $e_q$ is moved into any vacant room in timeslot $j_1$. If such changes are seen to violate any of the hard constraints, they are rejected immediately; else they are kept and the new solution is evaluated according to Equation (8.14).

$N_2$:    This operates in the same manner as $N_1$. However, when seeking to insert an event into a timeslot, if no vacant, suitable room is available, a maximum matching algorithm is also executed to determine if a valid room allocation of the events can be found. A similar operator was used by Cambazard et al. (2008) in their winning competition entry.

$N_3$:    This is an extension of $N_2$. Specifically, if the proposed move in $N_2$ will result in a violation of Constraint (8.6), then a Kempe chain interchange is attempted (see Definition 3.1). An example of this process is shown in Figure 8.3(a). Imagine in this case that we have chosen to swap the events $e_5 \in S_i$ and $e_{10} \in S_j$. However, doing so would violate Constraint (8.6) because events $e_5$ and $e_{11}$ conflict but would now both be assigned to timeslot $S_j$. In this case we therefore construct the Kempe chain $\text{KEMPE}(e_5, i, j) = \{e_5, e_{10}, e_{11}\}$ which, when interchanged, guarantees the preservation of Constraint (8.6), as shown in Figure 8.3(b). Observe that this neighbourhood operator also includes pair swaps (see Definition 3.2)—for example, if we were to select events $e_4 \in S_i$ and $e_8 \in S_j$ from Figure 8.3(a).

Note, however, that as with the previous neighbourhood operators, applications of $N_3$ may not preserve the satisfaction of the remaining hard constraints. Such moves will again need to be rejected if this is the case.

$N_4$:    This operator extends $N_3$ by using the idea of *double* Kempe chains, originally proposed by Lü and Hao (2010b). In many cases, a proposed Kempe chain interchange will be rejected because it will violate Constraint (8.10): that is, suitable rooms will not be available for all of the events proposed for assignment to a particular timeslot. For example, in Figure 8.3(a) the proposed Kempe interchange involving events $\{e_1, e_2, e_3, e_6, e_7\}$ is guaranteed to violate Constraint (8.10) because it will result in too many events in timeslot $S_j$ for a feasible matching to be possible. However, applying a second Kempe chain interchange at the same time may result in feasibility being maintained, as illustrated in Figure 8.3(c).

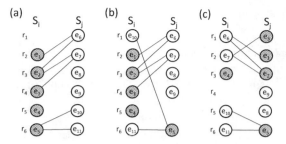

**Fig. 8.3** Example moves using $N_3$ and $N_4$. Here, edges exist between pairs of vertices (events) $e_l, e_q$ if and only if $C_{lq} = 1$. Part (a) shows two timeslots containing two Kempe chains, $\{e_1, e_2, e_3, e_6, e_7\}$ and $\{e_5, e_{10}, e_{11}\}$; part (b) shows a result of interchanging the latter chain; part (c) shows the result of interchanging both chains. Note that room allocations are determined via a matching algorithm and may therefore alter during an interchange

In this operator, if a proposed single Kempe chain interchange is seen to violate Constraint (8.10) only, then a random vertex from one of the two timeslots, but from outside this chain, is randomly selected, and a second Kempe chain is formed from it. If the proposed interchange of both Kempe chains does not violate any of the hard constraints, then the move can be performed and the new solution can be evaluated according to Equation (8.14) as before.

$N_5$: Finally, $N_5$ defines a *multi*-Kempe chain operator. This generalises $N_4$ in that if a proposed double Kempe chain interchange is seen to violate Constraint (8.10) only, then triple Kempe chains, quadruple Kempe chains, and so on, can also be investigated in the same manner. Note that when constructing these multiple Kempe chains, a violation of any of the constraints (8.7)–(8.9) allows us to reject the move immediately. However, if only Constraint (8.10) continues to be violated, then eventually the considered Kempe chains will contain all events in both timeslots, in which case the move becomes equivalent to swapping the contents of the two timeslots. Trivially, in such a move Constraint (8.10) is guaranteed to be satisfied.

From the above descriptions it is clear that each successive neighbourhood operator requires more computation than its predecessor. Each operator also generalises its predecessor—that is, $N_1(\mathcal{S}) \subseteq N_2(\mathcal{S}) \subseteq \ldots \subseteq N_5(\mathcal{S})$, $\forall \mathcal{S} \in \mathbb{S}$. From the perspective of the graph $G = (\mathbb{S}, E)$ defined above, this implies a greater connectivity of the solution space since $E_1 \subseteq E_2 \subseteq \ldots \subseteq E_5$ (where $E_i = \{\{\mathcal{S}, \mathcal{S}'\} : \mathcal{S}' \in N_i(\mathcal{S})\}$ for $i = 1, \ldots, 5$). Note though that the set of vertices (solutions) $\mathbb{S}$ remains the same under the different operators.

Finally, it is also worth mentioning that each of the above operators only ever alters the contents of two timeslots in any particular move. In practice, this means that we only need to consider the particular days and students affected by the move when reevaluating the solution according to Equation (8.14). This allows considerable speed-up of the algorithm.

### 8.5.3 Dummy Rooms

An additional opportunity for altering the connectivity of the underlying solution space with this problem is through the use of *dummy rooms*. A dummy room is an extra room made available in all timeslots and defined as suitable for all events (i.e., it has an infinite seating capacity and possesses all available room features). Dummy rooms can be used with any of the previous neighbourhood operators, and multiple dummy rooms can also be applied if necessary. We therefore use the notation $N_i^{(j)}$ to denote the use of neighbourhood $N_i$ using $j$ dummy rooms (giving $|r| + j$ rooms in total). Similarly, we can use the notation $\mathbb{S}^{(j)}$ to denote the space of all solutions that obey all hard constraints, and where $j$ dummy rooms are available. (For brevity, where a superscript is not used, we assume no dummy room is being used.) Dummy rooms can be viewed as a type of "placeholder" that are used to contain events not currently assigned to the "real" timetable. Transferring events in and out of dummy rooms might therefore be seen as similar to moving events in and out of the timetable.

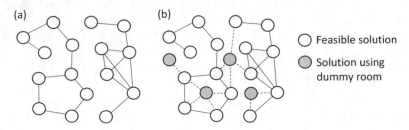

**Fig. 8.4** Graphs depicting the connectivity of a solution space with (a) no dummy rooms, and (b) one or more dummy rooms

Note that the use of dummy rooms increases the size of the set $\mathcal{M}$, making Constraint (8.10) easier to satisfy. This leads to the situation depicted in Figure 8.4. Here, we observe that the presence of dummy rooms increases the number of vertices/solutions (i.e., $\mathbb{S}^{(j)} \subseteq \mathbb{S}^{(j+1)}, \forall j \geq 0$), with extra edges (dotted in the figure) being created between some of the original vertices and new vertices. As depicted, this could also allow previously disjoint components to become connected.

Because they do not form part of the original problem, at the end of the optimisation process all dummy rooms will need to be removed, meaning that any events assigned to these will contribute to the DTF measure. Because this is undesirable, in our case we attempt to discourage the assignment of events to the dummy rooms during evaluation by considering all events assigned to a dummy room as unplaced. We then use the cost function $w \times \text{DTF} + \text{SCP}$, where $w$ is a weighting coefficient that will need to be set by the user. Additionally, when employing the maximum matching algorithm it also makes sense to ensure that the dummy room is only used when necessary—that is, if a feasible matching can be achieved without using dummy rooms, then this is the one that will be used.

### 8.5.4 Estimating Solution Space Connectivity

We have now seen various neighbourhood operators for this problem and made some observations on the connectivity of their underlying solution spaces, defined by $G$. Unfortunately however, it is very difficult to gain a complete understanding of $G$'s connectivity because it is simply too large to compute. In particular, we are unlikely to be able to confirm whether $G$ is connected or not, which would be useful information if we wanted to know whether an optimal solution could be reached from any other solution within the solution space.

One way to gain an indication of $G$'s connectivity is to make use of what we call the *feasibility ratio*. This is defined as the proportion of proposed neighbourhood moves that are seen to not violate any of the hard constraints (i.e., that maintain validity/feasibility). A lower feasibility ratio suggests a lower connectivity in $G$ because, on average, more potential moves will be seen to violate a hard constraint from a particular solution, making movements within the solution space more restricted. A higher feasibility ratio will suggest a greater level of connectivity.

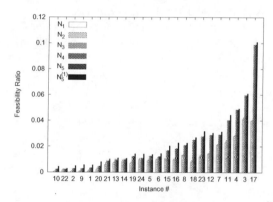

**Fig. 8.5** Feasibility ratio for neighbourhood operators $N_1 \ldots N_5$ and also $N_5^{(1)}$ for all 24 problem instances

Figure 8.5 displays the feasibility ratios for neighbourhood operators $N_1 \ldots N_5$ and also $N_5^{(1)}$ for all 24 problem instances. These mean figures were found by performing random walks of 50,000 feasible-preserving moves from a sample of 20 feasible solutions per instance (produced via Stage 1). As expected, we see that the feasibility ratios increase for each successive neighbourhood operator, though the differences between $N_3$, $N_4$, and $N_5$ appear to be only marginal. We also observe quite a large range across the instances, with instance #10 appearing to exhibit the least connected solution space (with feasibility ratios ranging from just 0.0005 ($N_1$) to 0.004 ($N_5^{(1)}$)), and instance #17 having the highest levels of connectivity (0.04 ($N_1$) to 0.10 ($N_5^{(1)}$)). Standard deviations from these samples range between

0.000018 ($N_2$, #20) and 0.000806 ($N_1$, #23). These observations will help to explain the results in the following sections.

## 8.6 Experimental Results

### 8.6.1 Effect of Neighbourhood Operators

We now examine the ability of each neighbourhood operator to reduce the soft constraint cost within the time limit specified by the competition benchmarking program (minus the time used for Stage 1). We also consider the effects of altering the end temperature of simulated annealing $T_{end}$, which is the only run-time parameter required in this stage. To measure performance, we compare our results to those achieved by the five finalists of the 2007 competition using the competition's ranking system. This involves calculating a "ranking score" for each algorithm, which is derived as follows.

Given $x$ algorithms and a single problem instance, each algorithm is executed $y$ times, giving $xy$ results. These results are then ranked from 1 to $xy$, with ties receiving a rank equal to the average of the ranks they span. The mean of the ranks assigned to each algorithm is then calculated, giving the respective rank scores for the $x$ algorithms on this instance. This process is then repeated on all instances, and the mean of all ranking scores for each algorithm is taken as its overall ranking score. A worked example of this process is shown in Table 8.3. We see that the best ranking score achievable for an algorithm on a particular instance is $(y+1)/2$, in which case its $y$ results are better than all of the other algorithms' (as is the case with Algorithm A on instance #3 in the table). The worst possible ranking score, $(x-1)y+(y+1)/2$, has occurred with Algorithm C with instances #1, #2, and #3 in the table.

**Table 8.3** Worked example of how rank scores are calculated using $y = 2$ runs of $x = 3$ algorithms on three problem instances. Results of each run are given by the DTF (in parentheses) and the SCC. Here, Algorithm A is deemed the winner and C the loser

| Instance | Results #1 | | #2 | | #3 | | Ranks #1 | #2 | #3 | Rank Scores #1 | #2 | #3 | Mean |
|---|---|---|---|---|---|---|---|---|---|---|---|---|---|
| Run | 1 | 2 | 1 | 2 | 1 | 2 | | | | | | | |
| Alg. A | (0) 0 | (0) 10 | (0) 1 | (0) 5 | (0) 0 | (0) 2 | 1, 3 | 1.5, 3 | 1, 2 | 2 | 2.25 | 1.5 | 1.92 |
| Alg. B | (0) 5 | (0) 17 | (0) 1 | (0) 8 | (0) 8 | (0) 11 | 2, 4 | 1.5, 4 | 3, 4 | 3 | 2.75 | 3.5 | 3.08 |
| Alg. C | (9) 3 | (0) 19 | (0) 18 | (0) 16 | (0) 12 | (4) 0 | 6, 5 | 6, 5 | 5, 6 | 5.5 | 5.5 | 5.5 | 5.50 |

A full breakdown of the results and ranking scores of the five competition finalists can be found on the official website of ITC2007, where $x = 5$ and $y = 10$.[4] In our case, we added results from ten runs of our algorithm to these published results,

---

[4] www.cs.qub.ac.uk/itc2007/

giving $x = 6$ and $y = 10$. A summary of the resultant ranking scores achieved by our algorithm with each neighbourhood operator over a range of different settings for $T_{end}$ is given in Figure 8.6. The shaded area of the figure indicates those settings where our algorithm would have won the competition (i.e., that have achieved a lower ranking score than the other five entries).

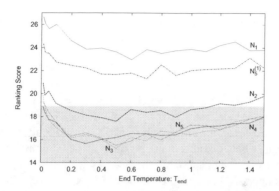

**Fig. 8.6** Ranking scores achieved by the different neighbourhood operators using different end temperatures. The shaded area indicates the results that would have won the competition

Figure 8.6 shows a clear difference in the performance of neighbourhood operators $N_1$ and $N_2$, illustrating the importance of the extra solution space connectivity provided by the maximum matching algorithm. Similarly, the results of $N_3$, $N_4$ and $N_5$ are better still, outperforming $N_1$ and $N_2$ across all the values of $T_{end}$ tested. However, there is very little difference between the performance of $N_3$, $N_4$ and $N_5$ themselves presumably due to the fact that, for these particular problem instances, the behaviour and therefore feasibility ratios of these operators are very similar (as seen in Figure 8.5). Moreover, we find that the extra expense of $N_5$ over $N_3$ and $N_4$ appears to have minimal effect, with $N_5$ producing less than 0.5% fewer Markov chains than $N_3$ over the course of the run on average. Of course, such similarities will not always be the case—they merely seem to be occurring with these particular problem instances because, in most cases, hard constraints are being broken (and the move rejected) before the inspection of more than one Kempe chain is deemed necessary.

Figure 8.6 also indicates that using dummy rooms does not seem to improve results across the instances. In initial experiments we tested the use of one and two dummy rooms along with a range of different values for the weighting coefficient $w \in \{1, 2, 5, 10, 20, 200, \infty\}$ (some of these values were chosen due to their use in existing algorithms that employ weighted sum functions with this problem (Ceschia et al., 2012; Nothegger et al., 2012; van den Broek and Hurkens, 2012)). Figure 8.6 reports the best of these: one dummy room with $w = 2$. For higher values of $w$, results were found to be inferior because the additional solutions in the solution space (shaded vertices in Figure 8.4) would still be evaluated by the algorithm, but nearly always rejected due to their high cost. On the other hand, using the setting $w = 1$ means that the penalty of assigning an event to a dummy room will be equal to the penalty of assigning the event to the last timeslot of a day (soft constraint SC1), meaning there is little distinction between the cost of infeasibility and the cost of soft constraint violations.

Note that we might consider the use of no dummy rooms as similar to using $w \approx \infty$, in that the algorithm will be unable to accept moves that involve moving an event into a dummy room (i.e., introducing infeasibility to the timetable). However, the difference is that when using dummy rooms, such moves will still be evaluated by the algorithm before being rejected while, when not dummy rooms, these unnecessary evaluations will not take place, saving significant amounts of time during the course of a run.

As mentioned, a setting of $w = 2$, which seems the best compromise between these extremes, still produces inferior results on average compared to when using no dummy rooms. However, for problem instance #10 we found the opposite to be true, with significantly better results being produced when dummy rooms *are* used. From Figure 8.5, we observe that #10 has the lowest feasibility ratio of all instances, and so the extra connectivity provided by the dummy rooms seems to be aiding the search in this case. On the other hand, the existence of a perfect solution here could mean that, while optimising the SCC, the search might also be being simultaneously guided towards regions of the solution space that are feasible. This matter will be discussed further in Section 8.7.

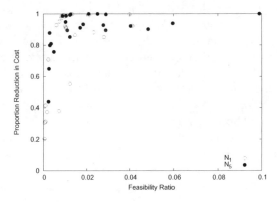

**Fig. 8.7** Scatter plot showing the relationship between feasibility ratio and reduction in cost for the 24 competition instances, using neighbourhood operators $N_1$ and $N_5$

Finally, Figure 8.7 illustrates for the 24 problem instances the relationship between the feasibility ratio and the proportion by which the SCC is reduced by the SA algorithm for two contrasting neighbourhood operators $N_1$ and $N_5$. We see that the points for $N_5$ are shifted upwards and rightwards compared to $N_1$, illustrating the larger feasibility ratios and higher performance of the operator. The general pattern in the figure suggests that higher feasibility ratios allow large decreases in cost during a run, while lower feasibility ratios can result in both large or small decreases, depending on the instance. Thus, while there is some relationship between the two variables, it seems that other factors also have an impact here, including the size and shape of the cost landscape and the amount of computation needed for each application of the evaluation function.

## 8.6.2 Comparison to Published Results

In our next set of experiments, we compare the performance of our algorithm to the best results that were reported in the literature in the five years following the 2007 competition. Table 8.4 gives a breakdown of the results achieved by our method using $T_{end} = 0.5$ compared to the approaches of van den Broek and Hurkens (2012), Cambazard et al. (2012), and Nothegger et al. (2012). Note that the latter two papers only list results for the first 16 instances. In this table, all statistics are calculated from 100 runs on each instance, with the exception of van den Broek and Hurkens, whose algorithm is deterministic. All results were achieved strictly within the time limits specified by the competition benchmark program. Our experiments were performed using $N_3$, $N_4$, and $N_5$, though no significant difference was observed between the three operators' best, mean, or worst results. Consequently we only present the results for one of these.[5]

Table 8.4 shows that, using $N_4$, perfect solutions have been achieved by our method in 17 of the 24 problem instances. A comparison to the 16 results reported by Cambazard et al. (2012) indicates that our method's best, mean, and worst results are also significantly better than their corresponding results. Similarly, our best, mean, and worst results are all seen to outperform the results of van den Broek and Hurkens (2012). Finally, no significant difference is observed between the best and mean results of our method compared to Nothegger et al. (2012); however, unlike our algorithm they have failed to achieve feasibility in a number of cases.

## 8.6.3 Differing Time Limits

In our final set of experiments we look at the effects of using differing time limits with our algorithm. Until this point, experiments have been performed according to the time limit specified by the competition benchmark program; however, it is pertinent to ask whether the less expensive neighbourhood operators are actually more suitable when shorter time limits are used, and whether further improvements can be achieved when the time limit is extended. In Figure 8.8 we show the relative performance of operators $N_1$, $N_2$, $N_3$, and $N_5$ using time limits of between 1 and 600 seconds, signifying very fast and very slow coolings respectively. ($N_4$ is omit-

---

[5] For pairwise comparisons, Related Samples Wilcoxon Signed Rank Tests were used; for other comparisons Friedman Tests were used (significance level 0.05).

**Table 8.4** Results from the literature, taken from samples of 100 runs per instance. Figures indicate the SCC achieved at the cut-off point defined by the competition benchmarking program. Numbers in parentheses indicate the % of runs where feasibility was found. No parentheses indicates that feasibility was achieved in all runs

| # | Our method using $N_4$ | | | Cambazard[a] | | | van den Broek[b] | Nothegger[c] | |
|---|---|---|---|---|---|---|---|---|---|
| | Best | Mean | Worst | Best | Mean | Worst | Result | Best | Mean |
| 1 | 0 | 377.0 | 833 | 15 | 547 | 1072 | 1636 | 0 | (54) 613 |
| 2 | 0 | 382.2 | 1934 | 9 | 403 | 1254 | 1634 | 0 | (59) 556 |
| 3 | 122 | 181.8 | 240 | 174 | 254 | 465 | 355 | 110 | 680 |
| 4 | 18 | 319.4 | 444 | 249 | 361 | 666 | 644 | 53 | 580 |
| 5 | 0 | 7.5 | 60 | 0 | 26 | 154 | 525 | 13 | 92 |
| 6 | 0 | 22.8 | 229 | 0 | 16 | 133 | 640 | 0 | (95) 212 |
| 7 | 0 | 5.5 | 11 | 1 | 8 | 32 | 0 | 0 | 4 |
| 8 | 0 | 0.6 | 59 | 0 | 0 | 0 | 241 | 0 | 61 |
| 9 | 0 | 514.4 | 1751 | 29 | 1167 | 1902 | 1889 | 0 | (85) 202 |
| 10 | 0 | 1202.4 | 2215 | 2 | (89) 1297 | 2637 | 1677 | 0 | 4 |
| 11 | 48 | 202.6 | 358 | 178 | 361 | 496 | 615 | 143 | (99) 774 |
| 12 | 0 | 340.2 | 583 | 14 | 380 | 676 | 528 | 0 | (86) 538 |
| 13 | 0 | 79.0 | 269 | 0 | 135 | 425 | 485 | 5 | (94) 360 |
| 14 | 0 | 0.5 | 7 | 0 | 15 | 139 | 739 | 0 | 41 |
| 15 | 0 | 139.9 | 325 | 0 | 47 | 294 | 330 | 0 | 29 |
| 16 | 0 | 105.2 | 223 | 1 | 58 | 245 | 260 | 0 | 101 |
| 17 | 0 | 0.1 | 3 | - | - | - | 35 | - | - |
| 18 | 0 | 2.2 | 57 | - | - | - | 503 | - | - |
| 19 | 0 | 346.1 | 1222 | - | - | - | 963 | - | - |
| 20 | 557 | 724.5 | 881 | - | - | - | 1229 | - | - |
| 21 | 1 | 32.1 | 159 | - | - | - | 670 | - | - |
| 22 | 4 | 1790.1 | 2280 | - | - | - | 1956 | - | - |
| 23 | 0 | 514.1 | 1178 | - | - | - | 2368 | - | - |
| 24 | 18 | 328.2 | 818 | - | - | - | 945 | - | - |

[a] SA-colouring method (Cambazard et al. (2012), p. 122).

[b] Deterministic IP-based heuristic (one result per instance) (van den Broek and Hurkens (2012), p. 451).

[c] Serial ACO algorithm (Nothegger et al. (2012), p. 334).

ted here due to its close similarity with $N_3$ and $N_5$'s results.) We see that even for very short time limits of less than five seconds, the more expensive neighbourhoods consistently produce superior solutions across the instances. We also see that when the time limit is extended beyond the benchmark and up to 600 seconds, the mean reduction in the soft cost rises from 89.1% to 94.6% (under $N_5$), indicating that superior results can also be gained with additional computing resources. This latter observation is consistent with that of Nothegger et al. (2012), who were also able to improve the results of their algorithm, in their case via parallelisation.

## 8.7 Chapter Summary and Discussion

This chapter has considered the problem of constructing university timetables—a problem that can often involve a multitude of different constraints and requirements. Despite these variations, like the case studies seen in the previous two chapters,

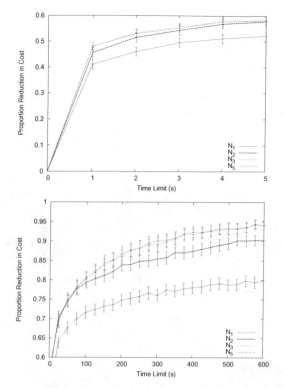

**Fig. 8.8** Proportion decrease in SCC using differing time limits and differing neighbourhood operators. All points are taken from an average of ten runs on each problem instance (i.e., 240 runs). Error bars represent one standard error each side of the mean

this problem usually contains an underlying graph colouring problem from which powerful algorithmic operators can then be derived.

Using this link with graph colouring, we have proposed a robust two-stage algorithm for a well-known NP-hard timetabling formulation known as the post enrolment-based timetabling problem. Stage 1 of this algorithm has proven to be very successful for finding feasibility with the considered problem instances with regard to both success rate and computation time. For Stage 2 we have then focussed on issues surrounding the connectivity of the solution space, and have seen that results generally improve when this connectivity is increased.

It is noticeable that many successful algorithms for the post enrolment-based timetabling problem have used simulated annealing as their main mechanism for reducing the number of soft constraint violations (Kostuch, 2005; Chiarandini et al., 2006; Cambazard et al., 2012). An alternative to this metaheuristic is, of course, tabu search; however, it does not seem to have fared as favourably with this problem formulation in practice. A contributing factor behind this lack of performance could be due to the observations made in this chapter—that a decreased connectivity of the solution space tends to lead to fewer gains being made during the optimisation process. To illustrate this, consider the situation shown in Figure 8.9 where a so-lution space and neighbourhood operator is again defined as a graph $G = (\mathbb{S}, E)$. In the top example, we show the effect of performing a neighbourhood move (i.e., changing the incumbent solution) with simulated annealing. In particular, we see that the connectivity of $G$ does not change (though the probabilities of traversing the edges may change if the temperature parameter is subsequently updated). On the other hand, when the same move is performed using tabu search, a number of edges in $G$, including $\{S, S'\}$, will be made tabu for a number of iterations, effec-tively removing them from the graph for a period of time dictated by the tabu tenure. The exact edges that will be made tabu depends on the structure of the tabu list, and in typical applications, when an event $e_i$ has been moved from timeslot $S_j$ to a new timeslot, *all* moves that involve moving $e_i$ back into $S_j$ will be made tabu. While the use of tabu moves helps to prevent cycling (which may regularly occur with SA), it therefore also has the effect of further reducing the connectivity of $G$. Over the course of a run, the cumulative effects of this phenomenon may put tabu search at a disadvantage for these particular problems.

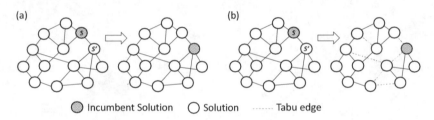

**Fig. 8.9** Illustration of the effects of performing a neighbourhood move using (a) simulated an-nealing, and (b) and tabu search

In this chapter we have also noted that an alternative approach to a two-stage algorithm is to use a one-stage optimisation algorithm in which the satisfaction of both hard and soft constraints is attempted simultaneously, as with the methods of Ceschia et al. (2012) and Nothegger et al. (2012). As we have seen, despite the favourable performance of our two-stage algorithm overall, it does seem to struggle in comparison to these approaches for a small number of problem instances, par-ticularly #10 and #22. According to Figure 8.5, these instances exhibit the lowest feasibility ratios with our operators, seemingly suggesting that freedom of move-ment in the solution spaces is too restricted to allow adequate optimisation of the objective function.

On the other hand, it is also possible that the algorithms of Ceschia et al. (2012) and Nothegger et al. (2012) are being aided by the fact that perfect solutions to the 24 competition instances are known to exist—a feature that is unlikely to occur in real-world problem instances. For example, as mentioned in Section 8.3, Ceschia et al.'s algorithm is reported to produce its best results when optimisation is performed us-ing an objective function in which hard and soft constraint violations are given equal weights. However, it could be that, by moving towards solutions with low SCCs, the search could also inadvertently be moving towards feasible regions of the solution space, simultaneously helping to satisfy the hard constraints along the way. This hypothesis was tested by Lewis and Thompson (2015), who compared the algorithm of this chapter to Ceschia et al.'s using a different suite of timetabling problems for which the existence of perfect solutions is not always known.[6] The results of their tests strongly support this hypothesis: for the 23 instances of this suite with no known perfect solution, the two-stage algorithm outperformed Ceschia et al.'s in 21 cases (91.3%), with stark differences in results. On the other hand, with the remaining 17 instances, Ceschia et al.'s approach produced better results in 12.5 cases (73.5%), suggesting that the existence of perfect solutions indeed benefits the algorithm.

As mentioned earlier, all of the problem instances used in this chapter are available online at www.cs.qub.ac.uk/itc2007/. In addition, a full listing of this chapter's results, together with C++ source code of the two-stage algorithm is available at www.rhydlewis.eu/resources/ttCodeResults.zip.

---

[6] These can be downloaded from www.rhydlewis.eu/hardTT.

# Appendix A
# Computing Resources

## A.1 Algorithm User Guide

This section contains instructions on how to compile and use the implementations of the algorithms described in Chapters 2 and 4 of this book. These can be downloaded directly from:

> http://rhydlewis.eu/resources/gCol.zip

Once downloaded and unzipped, we see that the directory contains a number of sub-directories. Each algorithm is contained within its own subdirectory. Specifically, these are:

- `AntCol` The Ant Colony Optimisation-based algorithm for graph colouring (see Section 4.1.4).
- `BacktrackingDSatur` The Backtracking algorithm based on the DSatur heuristic (see Section 4.1.6).
- `DSatur` The DSATUR algorithm (see Section 2.3).
- `HillClimber` The hill-climbing algorithm (see Section 4.1.5).
- `HybridEA` The hybrid evolutionary algorithm (see Section 4.1.3).
- `PartialColAndTabuCol` The PARTIALCOL and TABUCOL algorithms (see Sections 4.1.1 and 4.1.2 respectively).
- `RLF` The recursive largest first (RLF) algorithm (see Section 2.4).
- `SimpleGreedy` The GREEDY algorithm, using a random permutation of the vertices (see Section 2.1).

All of these algorithms are programmed in C++. They have been successfully compiled in Windows using Microsoft Visual Studio 2010 and in Linux using the GNU compiler g++. Instructions on how to do this now follow.

© Springer International Publishing Switzerland 2016

R.M.R. Lewis, *A Guide to Graph Colouring*,

DOI 10.1007/978-3-319-25730-3

### A.1.1 Compilation in Microsoft Visual Studio

To compile and execute using Microsoft Visual Studio the following steps can be taken:

1. Open Visual Studio and click **File**, then **New**, and then **Project from Existing Code**.
2. In the dialogue box, select **Visual C++** and click **Next**.
3. Select one of the subdirectories above, give the project a name, and click **Next**.
4. Finally, select **Console Application Project** for the project type, and then click **Finish**.

The source code for the chosen algorithm can then be viewed and executed from the Visual Studio application. Release mode should be used during compilation to make the programs execute at maximum speed.

### A.1.2 Compilation with g++

To compile the source code using g++, at the command line navigate to each sub-directory in turn and use the following command:

```
g++ *.cpp -O3 -o myProgram
```

By default this will create a new executable program called myProgram that can then be run from the command line (you should choose your own name here). The optimisation option -O3 ensures that the algorithms execute at maximum speed. Makefiles are also provided for compiling all algorithms in one go, if preferred.

### A.1.3 Usage

Once generated, the executable files (one per subdirectory) can be run from the command line. If the programs are called with no arguments, useful usage information is printed to the screen. For example, suppose we are using the executable file hillClimber. Running this program with no arguments from the command line gives the following output:

```
Hill Climbing Algorithm for Graph Colouring

USAGE:
<InputFile> (Required. File must be in DIMACS format)
-s <int>    (Stopping criteria expressed as number of constraint
             checks. Can be anything up to 9x10^18.
             DEFAULT = 100,000,000.)
-I <int>    (Number of iterations of local search per cycle.
             DEFAULT = 1000)
```

```
-r <int>     (Random seed. DEFAULT = 1)
-T <int>     (Target number of colours. Algorithm halts if this
             is reached. DEFAULT = 1.)
-v           (Verbosity. If present, output is sent to screen.
             If -v is repeated, more output is given.)
****
```

The input file should contain the graph colouring problem to be solved. This is the only mandatory argument. This must be in the DIMACS format, as described here:

> mat.gsia.cmu.edu/COLOR/general/ccformat.ps

For reference, an example input file called graph.txt is provided in each subdirectory.

The remaining arguments for each of the programs are optional and are allocated default values if left unspecified. Here are some example commands using the hillClimber executable:

> hillClimber graph.txt

This will execute the algorithm on the problem given in the file graph.txt, using the default of 1,000 iterations of local search per cycle and a random seed of 1. The algorithm will halt when 100,000,000 constraint checks have been performed. No output will be written to the screen.

Another example command is:

> hillClimber graph.txt -r 6 -T 50 -v -s 500000000000

This run will be similar to the previous one, but will use the random seed 6 and will halt either when 500,000,000,000 constraint checks have been performed, or when a feasible solution using 50 or fewer colours has been found. The presence of -v means that output will be written to the screen. Including -v more than once will increase the amount of output.

The arguments -r and -v are used with all of the algorithms supplied here. Similarly, -T and -s are used with all algorithms except for the single-parse constructive algorithms DSATUR, RLF and GREEDY. Descriptions of arguments particular to just one algorithm are found by typing the name of the program with no arguments, as described above. Interpretations of the run-time parameters for the various algorithms can be found by consulting the algorithm descriptions in this book.

## A.1.4 Output

When a run of any of the programs is completed, three files are created:

- cEffort.txt (computational effort),

- `tEffort.txt` (time effort), and
- `solution.txt`.

The first two specify how long (in terms of constraint checks and milliseconds respectively) solutions with certain numbers of colours took to produce during the last run. For example, we might get the following computational effort file:

| | |
|---|---|
| 40 | 126186 |
| 39 | 427143 |
| 38 | 835996 |
| 37 | 1187086 |
| 36 | 1714932 |
| 35 | 2685661 |
| 34 | 6849302 |
| 33 | X |

This file is interpreted as follows: The first feasible solution observed used 40 colours, and this took 126,186 constraint checks to achieve. A solution with 39 colours was then found after 427,143 constraint checks, and so on. To find a solution using 34 colours, a total of 6,849,302 constraint checks was required. Once a row with an X is encountered, this indicates that no further improvements were made: that is, no solution using fewer colours than that indicated in the previous row was achieved. Therefore, in this example, the best solution found used 34 colours. For consistency, the X is always present in a file, even if a specified target has been met.

The file `tEffort.txt` is interpreted in the same way as `cEffort.txt`, with the right hand column giving the time (in milliseconds) as opposed to the number of constraint checks. Both of these files are useful for analysing algorithm speed and performance. For example, the computational effort file above can be used to generate the following plot:

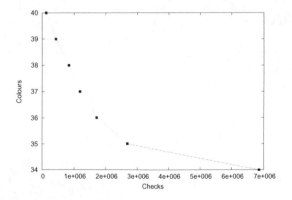

Finally, the file `solution.txt` contains the best feasible solution (i.e., the solution with fewest colours) that was achieved during the run. The first line of this file gives the number of vertices $n$, and the remaining $n$ lines then state the colour of each vertex, using labels $0, 1, 2, \ldots$.

For example, the following solution file

```
5
0
2
1
0
1
```

is interpreted as follows: There are five vertices; the first and forth vertices are assigned to colour 0, the third and fifth vertices are assigned to colour 1, and the second vertex is assigned to colour 2.

## A.2  Graph Colouring in Sage

Sage is specialised software that allows the exploration of many aspects of mathematics, including combinatorics, graph theory, algebra, calculus and number theory. It is both free to use and open source. To use Sage, commands can be typed into a notebook. Blocks of commands are then executed by typing Shift + Enter next to these commands, with output (if applicable) then being written back to the notebook.

Sage contains a whole host of elementary and specialised mathematical functions that are documented online at www.sagemath.org/doc/reference/. Of particular interest to us here is the functionality surrounding graph colouring and graph visualisations. A full description of the graph colouring library for Sage can be found at:

```
www.sagemath.org/doc/reference/graphs/sage/graphs/
     graph_coloring.html
```

The following text now shows some example commands from this library, together with the output that Sage produces. In our case, these commands have been typed into notebooks provided by the online tool at SageMathCloud. This tool allows the editing and execution of Sage notebooks through a web browser and can be freely accessed online at:

```
https://cloud.sagemath.com
```

The following pieces of code each represent an individual block of executable Sage commands. Any output produced by these commands are preceded by the ">>" symbol in the following text.

To begin, it is first necessary to specify the names of the libraries we intend to use in our Sage program. We therefore type:

```
from sage.graphs.graph_coloring import chromatic_number
from sage.graphs.graph_coloring import vertex_coloring
from sage.graphs.graph_coloring import number_of_n_colorings
from sage.graphs.graph_coloring import edge_coloring
```

which will allow us to access the various graph colouring functions used below.

We will now generate in Sage a small graph called G. In our case this graph has $n = 4$ vertices and $m = 5$ edges and is defined by the adjacency matrix

$$\mathbf{A} = \begin{pmatrix} 0 & 1 & 1 & 0 \\ 1 & 0 & 1 & 1 \\ 1 & 1 & 0 & 1 \\ 0 & 1 & 1 & 0 \end{pmatrix}$$

The first Sage command below defines this matrix. The next command then transfers this information into a graph called G. Finally G.show() draws this graph to the screen.

```
A = matrix([[0,1,1,0],[1,0,1,1],[1,1,0,1],[0,1,1,0]])
G = Graph(A)
G.show()
>>
```

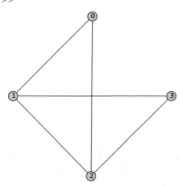

Note that, by default, Sage labels the vertices from $0, \ldots, n-1$ in this diagram as opposed to using indices $1, \ldots, n$.

We will now produce an optimal colouring of this graph. The algorithms that Sage uses to obtain these solutions are based on integer programming techniques (see Section 3.1.2). These are able to produce provably optimal solutions for small graphs such as this example; however, for larger graphs they are unlikely to return solutions in reasonable time. A colouring is produced via the following command (note the spelling of "coloring" as opposed to "colouring"):

```
vertex_coloring(G)
>> [[2], [1], [3, 0]]
```

The output produced by Sage tells us that G can be optimally coloured using three colours, with vertices 0 and 3 receiving the same colour. The solution returned by Sage is expressed as a partition of the vertices, which can be used to produce a visualisation of the colouring as follows:

```
S = vertex_coloring(G)
G.show(partition=S)
>>
```

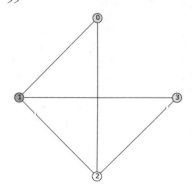

Here, the partition produced by `vertex_coloring(G)` is assigned to the variable S, which is then used as an additional argument in the `G.show` command to produce the above visualisation.

We can also use the `vertex_coloring` function to test if a graph is $k$-colourable. For example, to test whether G is 2-colourable, we get

```
vertex_coloring(G,2)
>> False
```

which tells us that a 2-colouring is not possible for this graph. On the other hand, if we seek to confirm whether G is 4-colourable, we get

```
vertex_coloring(G,4)
>> [[3, 0], [2], [1], []]
```

which tells us that one way of 4-colouring the graph G is to not use the fourth colour!

In addition to the above, commands are also available in Sage for determining the chromatic number

```
chromatic_number(G)
>> 3
```

and for calculating the number of different $k$-colourings. For example, with $k = 2$ we get

```
number_of_n_colorings(G,2)
>> 0
```

which is what we would expect since no 2-colouring of G exists. On the other hand, for $k = 3$ we get

```
number_of_n_colorings(G,3)
>> 6
```

telling us that there are six different ways of feasibly assigning three colours to G (readers are invited to confirm the correctness of this result themselves, by hand or otherwise).

In addition to vertex colouring, Sage also provides commands for calculating edge colourings of a graph (see Section 5.2). For example, continuing our use of the graph G from above, we can use the `edge_coloring()` command to get

```
edge_coloring(G)
>> [[(0, 1), (2, 3)], [(0, 2), (1, 3)], [(1, 2)]]
```

This tells us that the chromatic index of G is 4, with edges {0,1} and {2,3} being assigned to one colour, {0,2} and {1,3} being assigned to a second, and {1,2} being assigned to a third.

Sage also contains a collection of predefined graphs. This allows us to make use of common graph topologies without having to type adjacency matrices. A full list of these graphs is provided at:

```
www.sagemath.org/doc/reference/graphs/sage/graphs/
   graph_generators.html
```

For example, here are the commands for producing an optimal colouring of a do-
decahedral graph. In this case we have switched off vertex labelling to make the
illustration clearer:

```
G = graphs.DodecahedralGraph()
S = vertex_coloring(G)
G.show(partition=S, vertex_labels=False)
>>
```

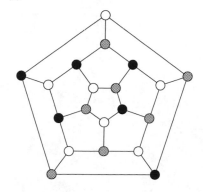

Here is an optimal colouring of the complete graph with ten vertices, $K_{10}$:

```
G = graphs.CompleteGraph(10)
S = vertex_coloring(G)
G.show(partition=S, vertex_labels=False)
>>
```

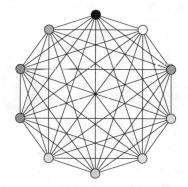

So-called lollypop graphs are defined by a path of $n_1$ vertices representing the
"stick" and a complete graph $K_{n_2}$ to represent the "head". Here is an example colouring using a graph with $n_1 = 4$ and $n_2 = 8$:

```
G = graphs.LollipopGraph(8,4)
S = vertex_coloring(G)
G.show(partition=S, vertex_labels=False)
>>
```

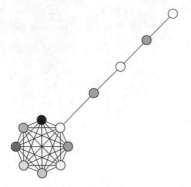

Our next graph, the "flower snark" is optimally coloured as follows:

```
G = graphs.FlowerSnark()
S = vertex_coloring(G)
G.show(partition=S, vertex_labels=False)
>>
```

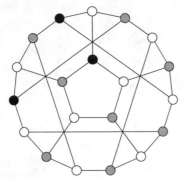

As we can see, this graph is 3-colourable. However, we can confirm that it is not planar using the following command:

```
G.is_planar()
>> False
```

Finally, Sage also allows us to define random graphs $G_{n,p}$ that have $n$ vertices and edge probabilities $p$ (see Definition 2.15). Here is an example with $n = 50$ and $p = 0.05$:

```
G = graphs.RandomGNP(50,0.05)
S = vertex_coloring(G)
G.show(partition=S, vertex_labels=False)
>>
```

It can be seen that this particular graph is 3-colourable, although the default layout of this graph is not very helpful, with the connected component on the left being very tightly clustered. If desired we can change this layout so that vertices are presented in a circle:

```
G.show(vertex_labels=False, layout='circular', partition=S)
>>
```

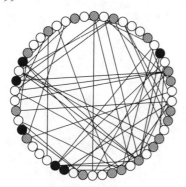

This, arguably, gives a clearer illustration of the graph.

## A.3  Graph Colouring with Commercial IP Software

The following code demonstrates how the graph colouring problem might be specified using integer programming methods and then solved using off-the-shelf optimisation software. This particular example, relating to the first IP model discussed
in Section 3.1.2, is coded in the Xpress-Mosel language, which comes as part of the
FICO Xpress Optimisation Suite. Comments in the code are preceded by exclama-
tion marks.

```
model GCOL
    !Gain access to the Xpress-Optimizer solver
    uses "mmxprs";

    !Define input file
    fopen("myGraph.txt",F_INPUT)

    !Define the integers used in the program
    declarations
        n,m,v1,v2: integer
    end-declarations

    !Read the num of vertices and edges from the input file
    read(n,m)
    writeln("n = ",n,",  m = ",m)

    !Declare the decision variable arrays
    declarations
        X: array(1..n,1..n) of mpvar
        Y: array(1..n) of mpvar
    end-declarations
    !And make all the variables binary
    forall (i in 1..n) do
        forall (j in 1..n) do
            X(i,j) is_binary
        end-do
        Y(i) is_binary
    end-do

    !Specify that each vertex should be assigned to exactly
    !one colour
    forall (i in 1..n) do
        sum(j in 1..n) X(i,j) = 1
    end-do

    !Now read in all of the edges and define the constraints
    write("E = {")
    forall (j in 1..m) do
        read(v1,v2)
        forall (i in 1..n) do
            X(v1,i) + X(v2,i) <= Y(i)
        end-do
        write("{",v1,",",v2,"}")
    end-do
```

```
writeln("}")

!Now specify the objective function
objfn := sum(i in 1..n) Y(i)

!Now run the model
writeln
writeln("Running model...")
minimize(objfn)
writeln("...Run ended")

!Finally write the output to the screen
writeln
writeln("Cost (number of colours) = ",getobjval)
writeln
writeln("X = ")
forall (i in 1..n) do
   forall (j in 1..n) do
      write(getsol(X(i,j))," ")
   end-do
   writeln
end-do
writeln
writeln("Y = ")
forall (j in 1..n) do
   write(getsol(Y(j))," ")
end-do
writeln
writeln
forall (i in 1..n) do
   write("c(v_",i,") = ")
   forall (j in 1..n) do
      if(getsol(X(i,j))=1) then
         writeln(j)
      end-if
   end-do
end-do

end-model
```

The above program starts by reading in a graph colouring problem from a text file (called myGraph.txt in this case). The objective function and constraints of the problem are then specified, before the optimisation process itself is invoked using the minimize(objfn) command. In this case the optimisation process is terminated only once a provably optimal solution has been found. However, other stopping conditions can also be specified if needed. Finally, the solution is written to the screen in a readable way.

Here is some example input that can be read in by the above program. The first two lines give the number of vertices and edges, $n$ and $m$, respectively. The $m$ edges then follow, one per line. This particular example corresponds to the graph shown in Figure 3.2.

```
8
12
1 2
1 3
1 4
2 5
2 6
2 8
3 4
3 7
4 7
5 8
6 8
7 8
```

On completion of the program, the following output is produced:

```
n = 8, m = 12
E = {{1,2}{1,3}{1,4}{2,5}{2,6}{2,8}{3,4}{3,7}{4,7}{5,8}{6,8}
     {7,8}}

Running model...
...Run ended

Cost (number of colours) = 3

X =
1 0 0 0 0 0 0 0
0 1 0 0 0 0 0 0
0 0 0 0 0 0 1 0
0 1 0 0 0 0 0 0
1 0 0 0 0 0 0 0
1 0 0 0 0 0 0 0
1 0 0 0 0 0 0 0
0 0 0 0 0 0 1 0

Y =
1 1 0 0 0 0 1 0

c(v_1) = 1
c(v_2) = 2
c(v_3) = 7
c(v_4) = 2
c(v_5) = 1
c(v_6) = 1
c(v_7) = 1
c(v_8) = 7
```

It can be seen that the cost of the solution (the number of colours being used) equals the chromatic number for this graph as expected. Note that, in this case, the colours with labels 1, 2 and 7 are being used to colour the vertices as opposed to 1, 2, and 3, which is permitted by this particular formulation.

## A.4 Useful Web Links

Here are some further web resources related to graph colouring. A page of resources maintained by Joseph Culberson featuring, most notably, a collection of problem generators and C code for the algorithms presented by Culberson and Luo (1996) can be found at:

`webdocs.cs.ualberta.ca/~joe/Coloring/`

An excellent bibliography on the graph coloring problem, maintained by Marco Chiarandini and Stefano Gualandi can also be found at:

`www.imada.sdu.dk/~marco/gcp/`

A large set of graph colouring problem instances has been collected by the Center for Discrete Mathematics and Theoretical Computer Science (DIMACS) as part of their DIMACS Implementation Challenge series. These can be downloaded at:

`mat.gsia.cmu.edu/COLOR/instances.html`

These problem instances have been used in a large number of graph colouring-based papers and are written in the DIMACS graph format, a specification of which can be found in the following (postscript) document:

`mat.gsia.cmu.edu/COLOR/general/ccformat.ps`

Note that these instances can also be viewed via a text editor. A summary of these instances, including their best known bounds, is maintained by Daniel Porumbel and is available at:

`www.info.univ-angers.fr/pub/porumbel/graphs/`

The fun graph colouring game *CoLoRaTiOn*, which is suitable for both adults and children, can be downloaded from:

`http://vispo.com/software/`

The goal in this game is to achieve a feasible colouring within in a certain number of moves. The difficulty of each puzzle depends on a number of factors, including its topology, whether you can see all of the edges, the number of vertices, and the number of available colours.

Finally, C++ code for the random Sudoku problem instance generator used in Section 5.4.1 of this book can be downloaded from:

`rhydlewis.eu/resources/sudokuGeneratorMetaheuristics`
   `.zip`

A Sudoku to graph colouring problem converter can also be found at:

`rhydlewis.eu/resources/sudokuToGCol.zip`

When compiled, this program reads in a single Sudoku problem (from a text file) and converts it into the equivalent graph colouring problem in the DIMACS format mentioned above.

# References

K. Aardel, S. van Hoesel, A. Koster, C. Mannino, and A. Sassano. Models and solution techniques for the frequency assignment problems. *4OR : Quarterly Journal of the Belgian, French and Italian Operations Research Societies*, 1(4): 1–40, 2002.

A. Anagnostopoulos, L. Michel, P. van Hentenryck, and Y. Vergados. A simulated annealing approach to the traveling tournament problem. *Journal of Scheduling*, 9(2):177–193, 2006.

I. Anderson. Kirkman and $GK_2n$. *Bulletin of the Institute of Combinatorics and its Applications*, 3:111–112, 1991.

K. Appel and W. Haken. Solution of the four color map problem. *Scientific American*, 4:108121, 1977a.

K. Appel and W. Haken. Every planar map is four colorable. Part I. Discharging. *Illinois Journal of Mathematics*, 21:429–490, 1977b.

K. Appel and W. Haken. Every planar map is four colorable. Part II. Reducibility. *Illinois Journal of Mathematics*, 21:491–567, 1977c.

C. Avanthay, A. Hertz, and N. Zufferey. A variable neighborhood search for graph coloring. *European Journal of Operations Research*, 151:379–388, 2003.

T. Bartsch, A. Drexl, and S. Kroger. Scheduling the professional soccer leagues of Austria and Germany. *Computers and Operations Research*, 33(7):1907–1937, 2006.

L. Beineke and J. Wilson, editors. *Topics in Chromatic Graph Theory*. Encyclopedia of Mathematics and its Applications (no. 156). Cambridge University Press, 2015.

C. Berge. Les problémes de coloration en théorie des graphes. *Publ. Inst. Stat. Univ. Paris*, 9:123–160, 1960,

C. Berge. *Graphs and Hypergraphs*. North-Holland, 1970.

D. Bertsekas. Auction algorithms for network flow problems: A tutorial introduction. *Computational Optimization and Applications*, 1:7–66, 1992.

I. Blöchliger and N. Zufferey. A graph coloring heuristic using partial solutions and a reactive tabu scheme. *Computers and Operations Research*, 35:960–975, 2008.

© Springer International Publishing Switzerland 2016
R.M.R. Lewis, *A Guide to Graph Colouring*,
DOI 10.1007/978-3-319-25730-3

B. Bollobás. The chromatic number of random graphs. *Combinatorica*, 8(1):49–55, 1988.

B. Bollobás. *Modern Graph Theory*. Springer, 1998.

W. Boyer, J. Myrvold. On the cutting edge: simplified $\mathcal{O}(n)$ planarity by edge addition. *Journal of Graph Algorithms and Applications*, 8:241–273, 2004.

D. Brélaz. New methods to color the vertices of a graph. *Commun. ACM*, 22(4): 251–256, 1979.

R. Brooks. On colouring the nodes of a network. *Mathematical Proceedings of the Cambridge Philosophical Society*, 37:194–197, 1941.

E. Burke and J. Newall. A multi-stage evolutionary algorithm for the timetable problem. *IEEE Transactions on Evolutionary Computation*, 3(1):63–74, 1999.

E. Burke, D. Elliman, and R. Weare. Specialised recombinative operators for timetabling problems. In *The Artificial Intelligence and Simulated Behaviour Workshop on Evolutionary Computing*, volume 993, pages 75–85. Springer, 1995.

E. Burke, D. Elliman, P. Ford, and R. Weare. Examination timetabling in British universities: A survey. In E Burke and P Ross, editors, *Practice and Theory of Automated Timetabling (PATAT) I*, volume 1153 of *LNCS*, pages 76–92. Springer, 1996.

H. Cambazard, E. Hebrard, B. O'Sullivan, and A. Papadopoulos. Local search and constraint programming for the post enrolment-based course timetabling problem. In E. Burke and M. Gendreau, editors, *Practice and Theory of Automated Timetabling (PATAT) VII*, 2008.

H. Cambazard, E. Hebrard, B. O'Sullivan, and A. Papadopoulos. Local search and constraint programming for the post enrolment-based timetabling problem. *Annals of Operational Research*, 194:111–135, 2012.

M. Caramia and P. Dell'Olmo. Solving the minimum weigted coloring problem. *Networks*, 38(2):88–101, 2001.

M. Carrasco and M. Pato. A multiobjective genetic algorithm for the class/teacher timetabling problem. In E. Burke and W. Erben, editors, *Practice and Theory of Automated Timetabling (PATAT) III*, volume 2079 of *LNCS*, pages 3–17. Springer, 2001.

F. Carroll and R. Lewis. The "engaged" interaction: Important considerations for the HCI design and development of a web application for solving a complex combinatorial optimization problem. *World Journal of Computer Application and Technology*, 1(3):75–82, 2013.

M. Carter, G. Laporte, and S. Y. Lee. Examination timetabling: Algorithmic strategies and applications. *Journal of the Operational Research Society*, 47:373–383, 1996.

S. Ceschia, L. Di Gaspero, and A. Schaerf. Design, engineering, and experimental analysis of a simulated annealing approach to the post-enrolment course timetabling problem. *Computers and Operational Research*, 39:1615–1624, 2012.

G. Chaitin. Register allocation and spilling via graph coloring. *SIGPLAN Not.*, 39 (4):66–74, 2004.

M. Chams, A. Hertz, and O. Dubuis. Some experiments with simulated annealing for coloring graphs. *European Journal of Operations Research*, 32:260–266, 1987.

P. Cheeseman, B. Kanefsky, and W. Taylor. Where the really hard problems are. In *Proceedings of IJCAI-91*, pages 331–337, 1991.

M. Chiarandini and T. Stützle. An application of iterated local search to graph coloring. *Proceedings of the Computational Symposium on Graph Coloring and it's Generalizations*, pages 112–125, 2002.

M. Chiarandini, M. Birattari, K. Socha, and O. Rossi-Doria. An effective hybrid algorithm for university course timetabling. *Journal of Scheduling*, 9(5):403–432, 2006. ISSN 1094-6136. doi: http://dx.doi.org/10.1007/s10951-006-8495-8.

M. Chudnovsky, N. Robertson, P. Seymour, and R. Thomas. The strong perfect graph theorem. *Annals of Mathematics*, 164(1):51–229, 2006.

E. Coffman, M. Garey, D. Johnson, and A. LaPaugh. Scheduling file transfers. *SIAM Journal of Computing*, 14(3):744–780, 1985.

C. Colbourn. The complexity of completing partial Latin squares. *Discrete Applied Mathematics*, 8(1):25–30, 1984.

E. Coll, G. Duran, and P. Moscato. A discussion on some design principles for efficient crossover operators for graph coloring problems. *Anais do XXVII Simposio Brasileiro de Pesquisa Operacional, Vitoria-Brazil*, 1995.

A. Colorni, M. Dorigo, and V Maniezzo. Metaheuristics for high-school timetabling. *Computational Optimization and Applications*, 9(3):277–298, 1997.

S. Cook. The complexity of theorem-proving procedures. In *Proceedings of the Third Annual ACM Symposium on Theory of Computing*, STOC '71, pages 151–158, New York, NY, USA, 1971. ACM. doi: 10.1145/800157.805047.

T. Cooper and J. Kingston. The complexity of timetable construction problems. In E. Burke and P. Ross, editors, *Practice and Theory of Automated Timetabling (PATAT) I*, volume 1153 of *LNCS*, pages 283–295. Springer, 1996.

T. Cormen, C. Leiserson, R. Rivest, and C. Stein. *Introduction to Algorithms, Third Edition*. The MIT Press, 2009.

D. Corne, P. Ross, and H. Fang. Evolving timetables. In L. Chambers, editor, *The Practical Handbook of Genetic Algorithms*, volume 1, pages 219–276. CRC Press, 1995.

P. Cote, T. Wong, and R. Sabourin. Application of a hybrid multi-objective evolutionary algorithm to the uncapacitated exam proximity problem. In E. Burke and M. Trick, editors, *Practice and Theory of Automated Timetabling (PATAT) V*, volume 3616 of *LNCS*, pages 294–312. Springer, 2005.

D. Cranston and L. Rabern. Brooks' theorem and beyond. *Journal of Graph Theory*, 2014. doi: 10.1002/jgt.21847.

J. Culberson and F. Luo. Exploring the k-colorable landscape with iterated greedy. *American Mathematical Society: Cliques, Coloring, and Satisfiability – Second DIMACS Implementation Challenge*, 26:245–284, 1996.

D. de Werra. Some models of graphs for scheduling sports competitions. *Discrete Applied Mathematics*, 21:47–65, 1988.

K. Deb, A. Pratap, S. Agarwal, and T. Meyarivan. A fast elitist multi-objective genetic algorithm: NSGA-II. *IEEE Transactions on Evolutionary Computation*, 6(2):182–197, 2000.

F. della Croce and D. Oliveri. Scheduling the Italian football league: an ILP-based approach. *Computers and Operations Research*, 33(7):1963–1974, 2006.

F. della Croce, R. Tadei, and P. Asioli. Scheduling a round-robin tennis tournament under courts and players unavailability constraints. *Annals of Operational Research*, 92:349–361, 1999.

M. Demange, D. de Werra, J. Monnot, and V. Paschos. Time slot scheduling of compatible jobs. *Journal of Scheduling*, 10:111–127, 2007.

L. Di Gaspero and A. Schaerf. Multi-neighbourhood local search with application to course timetabling. In E. Burke and P. De Causmaecker, editors, *Practice and Theory of Automated Timetabling (PATAT) IV*, volume 2740 of *LNCS*, pages 263–287. Springer, 2002.

L. Di Gaspero and A. Schaerf. Neighborhood portfolio approach for local search applied to timetabling problems. *Journal of Mathematical Modeling and Algorithms*, 5(1):65–89, 2006.

L Di Gaspero and A. Schaerf. A composite-neighborhood tabu search approach to the traveling tournament problem. *Journal of Heuristics*, 13(2):189–207, 2007.

J. Dinitz, D. Garnick, and B. McKay. There are 526,915,620 nonisomorphic one-factorizations of $K_{12}$. *Journal of Combinatorial Designs 2*, 2:273–285, 1994.

M. Dorigo, V. Maniezzo, and A. Colorni. The ant system: Optimisation by a colony of cooperating agents. *IEEE Trans. Syst. Man Cybern*, 26(1):29–41, 1996.

R. Dorne and J.-K. Hao. A new genetic local search algorithm for graph coloring. In A. Eiben, T. Back, M. Schoenauer, and H. Schwefel, editors, *Parallel Problem Solving from Nature (PPSN) V*, volume 1498 of *LNCS*, pages 745–754. Springer, 1998.

A. Dupont, A. Linhares, C. Artigues, D. Feillet, P. Michelon, and M. Vasquez. The dynamic frequency assignment problem. *European Journal of Operational Research*, 195:75–88, 2009.

K. Easton, G. Nemhauser, and M. Trick. The traveling tournament problem: description and benchmarks. In T. Walsh, editor, *Principles and Practice of Constraint Programming*, volume 2239 of *LNCS*, pages 580–585. Springer, 2001.

K. Easton, G. Nemhauser, and M. Trick. Solving the traveling tournament problem: A combined integer programming and constraint programming approach. In E. Burke and P. De Causmaecker, editors, *Practice and Theory of Automated Timetabling (PATAT) IV*, volume 2740 of *LNCS*, pages 100–109. Springer, 2003.

J. Egeblad and D. Pisinger. Heuristic approaches for the two- and three-dimensional knapsack packing problem. *Computers and Operational Research*, 36(4):1026–1049, 2009.

A. Eiben, J. van der Hauw, and J. van Hemert. Graph coloring with adaptive evolutionary algorithms. *Journal of Heuristics*, 4(1):25–46, 1998.

M. Elf, M. Junger, and G. Rinaldi. Minizing breaks by maximizing cuts. *Operations Research Letters*, 31(5):343–349, 2003.

E. Erben. A grouping genetic algorithm for graph colouring and exam timetabling. *Practice and Theory of Automated Timetabling (PATAT) III*, 2079:132–158, 2001.

P. Erdős. *Theory of Graphs and its Applications*, chapter Problem 9, page 159. Czech Acad. Sci. Publ., 1964.

B. Escoffier, J. Monnot, and V. Paschos. Weighted coloring: further complexity and approximability results. *Information Processing Letters*, 97(3):98–103, 2006.

E. Falkenauer. *Genetic Algorithms and Grouping Problems*. John Wiley and Sons, 1998.

I. Finocchi, A. Panconesi, and R. Silvestri. An experimental analysis of simple, distributed vertex colouring algorithms. *Algorithmica*, 41:1–23, 2005.

C. Fleurent and J. Ferland. Allocating games for the NHL using integer programming. *Operations Research*, 41(4):649–654, 1993.

C. Fleurent and J. Ferland. Genetic and hybrid algorithms for graph colouring. *Annals of Operational Research*, 63:437–461, 1996.

H. Furmańczyk. *Graph Colorings*, chapter Equitable Coloring of Graphs, pages 35–54. American Mathematical Society, 2004.

P. Galinier and J.-K. Hao. Hybrid evolutionary algorithms for graph coloring. *Journal of Combinatorial Optimization*, 3:379–397, 1999.

P. Galinier and A. Hertz. A survey of local search algorithms for graph coloring. *Computers and Operations Research*, 33:2547–2562, 2006.

M. Garey and D. Johnson. The complexity of near-optimal coloring. *Journal of the Association for Computing Machinery*, 23(1):43–49, 1976.

M. Garey and D. Johnson. *Computers and Intractability - A guide to NP-completeness*. W. H. Freeman and Company, San Francisco, first edition, 1979.

M. Garey, D. Johnson, and H. So. An application of graph coloring to printed circuit testing. *IEEE Transactions on Circuits and Systems*, CAS-23:591–599, 1976.

B. Gendron, A. Hertz, and P. St-Louis. On edge orienting methods for graph coloring. *Journal of Combinatorial Optimization*, 13(2):163–178, 2007.

F. Glover. Future paths for integer programming and links to artificial intelligence. *Computers and Operations Research*, 13(5):533–549, 1986.

A. Gyárfás and J. Lehel. On-line and first fit colourings of graphs. *Journal of Graph Theory*, 12:217–227, 1988.

A. Hajnal and E. Szemerédi. *Combinatorial Theory and it's Application*, chapter Proof of a Conjecture by P. Erdős, pages 601–623. North-Holland, 1970.

P. Hansen, M. Labbé, and D. Schindl. Set covering and packing formulations of graph coloring: Algorithms and first polyhedral results. *Discrete Optimization*, 6 (2):135 – 147, 2009.

R. Hassin and J. Monnot. The maximum saving partition problem. *Operations Research Letters*, 33:242–248, 2005

P. Heawood. Map-colour theorems. *Quarterly Journal of Mathematics*, 24:332–338, 1890.

M. Henz, T. Muller, and S. Theil. Global constraints for round robin tournament scheduling. *European Journal of Operational Research*, 153:92–101, 2004.

H. Hernández and C. Blum. FrogSim: Distributed graph coloring in wireless ad hoc networks. *Telecommunication Systems*, 55:211–223, 2014.

A. Hertz and D. de Werra. Using tabu search techniques for graph coloring. *Computing*, 39(4):345–351, 1987.

A. Hertz, M. Plumettaz, and N. Zufferey. Variable space search for graph coloring. *Discrete Applied Mathematics*, 156(13):2551–2560, 2008.

A. Herzberg and M. Murty. Sudoku squares and chromatic polynomials. *Notices of the AMS*, 54(6):708–717, 2007.

A. Hoffman. On eigenvalues and colorings of graphs. In *Graph Theory and Its Applications*, Proc. Adv. Sem., Math., page 7991, Research Center, Univ. of Wisconsin, Madison, WI, 1969, Academic Press, New York, 1970.

I. Holyer. The NP-completeness of edge-coloring. *SIAM Journal on Computing*, 10:718–720, 1981.

J. Hopcroft and R. Tarjan. Efficient planarity testing. *Journal of the Association for Computing Machinery*, 21:549–568, 1974.

R. Janczewski, M. Kubale, K. Manuszewski, and K. Piwakowski. The smallest hard-to-color graph for algorithm DSatur. *Discrete Mathematics*, 236:151–165, 2001.

S. Jat and S. Yang. A hybrid genetic algorithm and tabu search approach for post enrolment course timetabling. *Journal of Scheduling*, 14:617–637, 2011.

T. Jensen and B. Toft. *Graph Coloring Problems*. Wiley-Interscience, first edition, 1994.

A. Kanevsky. Finding all minimum-size separating vertex sets in a graph. *Networks*, 23:533–541, 1993.

M. Karp. *Complexity of Computer Computations*, chapter Reducibility Among Combinatorial Problems, pages 85–103. Plenum, New York, 1972.

M. Kearns, S. Suri, and N. Montfort. An experimental study of the coloring problem on human subject networks. *Science*, 313:824–827, 2006.

A. Kempe. On the geographical problem of the four colours. *American Journal of Mathematics*, 2:193–200, 1879.

G. Kendall, S. Knust, C. Ribeiro, and S. Urrutia. Scheduling in sports, an annotated bibliography. *Computers and Operations Research*, 37(1):1–19, 2010.

H. Kierstead and A. Kostochka. A short proof of the Hajnal-Szemerédi theorem on equitable coloring. *Combinatorics, Probability and Computing*, 17:265–270, 2008.

H. Kierstead and W. Trotter. An extremal problem in recursive combinatorics. *Congressus Numerantium*, 33:143–153, 1981.

H. Kierstead, A. Kostochka, M. Mydlarz, and E. Szemerédi. A fast algorithm for equitable graph coloring. *Combinatorica*, 30:217–224, 2010.

T. Kirkman. On a problem in combinations. *Cambridge Dublin Math Journal*, 2: 191–204, 1847.

S. Kirkpatrick, C. Gelatt, and M. Vecchi. Optimization by simulated annealing. *Science*, 220(4598):671–680, 1983.

V. Kolmogorov. Blossom V: A new implementation of a minimum cost perfect matching algorithm. *Mathematical Programming Computation*, 1(1):43–67, 2009.

D. König. Gráfok és alkalmazásuk a determinánsok és a halmazok elméletére. *Matematikai és Természettudományi Értestö*, 34:104–119, 1916.

S. Korman. *Combinatorial Optimization.*, chapter The Graph-Coloring Problem, pages 211–235. Wiley, New York, 1979.

P. Kostuch. The university course timetabling problem with a 3-phase approach. In E. Burke and M. Trick, editors, *Practice and Theory of Automated Timetabling (PATAT) V*, volume 3616 of *LNCS*, pages 109–125. Springer, 2005.

M. Kubale and B. Jackowski. A generalized implicit enumeration algorithm for graph coloring. *Commun. ACM*, 28(28):412–418, 1985.

K. Kuratowski. Sur le probleme des courbes gauches en topologie. *Fundamenta Mathematicae*, 15:271–283, 1930.

M. Laguna and R. Marti. A grasp for coloring sparse graphs. *Computational Optimization and Applications*, 19:165–78, 2001.

F. Leighton. A graph coloring algorithm for large scheduling problems. *Journal of Research of the National Bureau of Standards*, 84(6):489–506, 1979.

R. Lewis. A survey of metaheuristic-based techniques for university timetabling problems. *OR Spectrum*, 30(1):167–190, 2008.

R. Lewis. A general-purpose hill-climbing method for order independent minimum grouping problems: A case study in graph colouring and bin packing. *Computers and Operations Research*, 36(7):2295–2310, 2009.

R. Lewis. A time-dependent metaheuristic algorithm for post enrolment-based course timetabling. *Annals of Operational Research*, 194(1):273–289, 2012.

R. Lewis. *Springer Handbook of Computational Intelligence*, chapter Graph Coloring and Recombination, pages 1239–1254. Studies in Computational Intelligence. Springer, 2015.

R. Lewis and B. Paechter. Finding feasible timetables using group based operators. *IEEE Transactions on Evolutionary Computation*, 11(3):397–413, 2007.

R. Lewis and J. Thompson. On the application of graph colouring techniques in round-robin sports scheduling. *Computers and Operations Research*, 38(1):190–204, 2010.

R. Lewis and J. Thompson. Analysing the effects of solution space connectivity with an effective metaheuristic for the course timetabling problem. *European Journal of Operational Research*, 240:637–648, 2015.

R. Lewis, J. Thompson, C. Mumford, and J. Gillard. A wide-ranging computational comparison of high-performance graph colouring algorithms. *Computers and Operations Research*, 39(9):1933–1950, 2012.

A. Lim, B. Rodrigues, and X. Zhang. A simulated annealing and hill-climbing algorithm for the traveling tournament problem. *European Journal of Operational Research*, 174(3):1459–1478, 2006

L. Lovász, M. Saks, and W. Trotter. An on-line graph colouring algorithm with sublinear performance ratio. *Discrete Mathematics*, 75:319–325, 1989.

Z. Lü and J.-K Hao. A memetic algorithm for graph coloring. *European Journal of Operational Research*, 203(1):241 – 250, 2010a.

Z. Lü and J-K. Hao. Adaptive tabu search for course timetabling. *European Journal of Operational Research*, 200(1):235–244, 2010b.

D. MacKenzie. Graph theory uncovers the roots of perfection. *Science*, 38:297, 2002.

E. Malaguti, M. Monaci, and P. Toth. A metaheuristic approach for the vertex coloring problem. *INFORMS Journal on Computing*, 20(2):302–316, 2008.

E. Malaguti, M. Monaci, and P. Toth. Models and heuristic algorithms for a weighted vertex coloring problem. *Journal of Heuristics*, 15:503–526, 2009.

E. Malaguti, M. Monaci, and P. Toth. An exact approach for the vertex coloring problem. *Discrete Optimization*, 8(2):174–190, 2011.

B. McCollum, A. Schaerf, B.. Paechter, P. McMullan, R. Lewis, A. Parkes, L. Di Gaspero, R. Qu, and E. Burke. Setting the research agenda in automated timetabling: The second international timetabling competition. *INFORMS Journal on Computing*, 22(1):120–130, 2010.

C. McDiarmid and B. Reed. Channel assignment and weighted coloring. *Networks*, 36(2):114–117, 2000.

G. McGuire, B. Tugemann, and G. Civario. There is no 16-clue Sudoku: Solving the Sudoku minimum number of clues problem. *Computing Research Repository*, abs/1201.0749, 2012.

A. Mehrotra and M. Trick. A column generation approach for graph coloring. *INFORMS Journal on Computing*, 8(4):344–354, 1996.

A. Mehrotra and M. Trick. *Extending the Horizons: Advances in Computing, Optimization, and Decision Technologies. Operations Research/Computer Science Interfaces Series Volume 37*, chapter A Branch-and-Price Approach for Graph Multi-coloring, pages 15–29. Springer, 2007.

I. Méndez-Díaz and P. Zabala. A cutting plane algorithm for graph coloring. *Discrete Applied Mathematics*, 156:159–179, 2008.

W. Meyer. Equitable coloring. *American Mathematical Monthly*, 80:920–922, 1973.

J. Misra and D. Gries. A constructive proof of Vizing's theorem. *Information Processing Letters*, 41:131–133, 1992.

R. Miyashiro and T. Matsui. A polynomial-time algorithm to find an equitable home-away assignment. *Operations Research Letters*, 33:235–241, 2005.

R. Miyashiro and T. Matsui. Minimizing the carry-over effects value in a round-robin tournament. In E. Burke and H. Rudova, editors, *PATAT '06 Proceedings of the 6th International Conference on the Practice and Theory of Automated Timetabling*, pages 460–464, 2006a.

R. Miyashiro and T. Matsui. Semidefinite programming based approaches to the break minimization problem. *Computers and Operations Research*, 33(7):1975–1992, 2006b.

J. Moody and D. White. Structural cohesion and embeddedness: A hierarchical concept of social groups. *American Sociological Review*, 68(1):103–127, 2003.

C. Morgenstern and H. Shapiro. Coloration neighborhood structures for general graph coloring. In *Proceedings of the First Annual ACM-SIAM Symposium on Discrete Algorithms*, pages 226–235, San Francisco, California, USA, Society for Industrial and Applied Mathematics, 1990.

T. Müller and H. Rudova. Real life curriculum-based timetabling. In D Kjenstad, A. Riise, T. Nordlander, B. McCollum, and E. Burke, editors, *Practice and Theory of Automated Timetabling (PATAT 2012)*, pages 57–72, 2012.

C. Mumford. New order-based crossovers for the graph coloring problem. In *Parallel Problem Solving from Nature (PPSN) IX*, volume 4193 of *LNCS*, pages 880–889. Springer, 2006.

J. Munkres. Algorithms for the assignment and transportation problems. *Journal of the Society for Industrial and Applied Mathematics*, 5(1):32–38, 1957.

J. Mycielski. Sur le coloriage des graphes. *Colloquium Mathematicum*, 3:161–162, 1955.

G. Nemhauser and M. Trick. Scheduling a major college basketball conference. *Operations Research*, 46:1–8, 1998.

J. Nielsen. The need for web design standards. Online Article: www.nngroup.com/-articles/the-need-for-web-design-standards, September 2004.

C. Nothegger, A. Mayer, A. Chwatal, and G. Raidl. Solving the post enrolment course timetabling problem by ant colony optimization. *Annals of Operational Research*, 194:325–339, 2012.

L. Ouerfelli and H. Bouziri. Greedy algorithms for dynamic graph coloring. In *Proceedings of the International Conference on Communications, Computing and Control Applications (CCCA)*, pages 1–5, 2011. doi: 10.1109/CCCA.2011.6031437.

B. Paechter, R. Rankin, A. Cumming, and T. Fogarty. Timetabling the classes of an entire university with an evolutionary algorithm. In T. Baeck, A. Eiben, M. Schoenauer, and H. Schwefel, editors, *Parallel Problem Solving from Nature (PPSN) V*, volume 1498 of *LNCS*, pages 865–874. Springer, 1998.

L. Paquete and T. Stützle. An experimental investigation of iterated local search for coloring graphs. In S. Cagnoni, J. Gottlieb, E. Hart, M. Middendorf, and G. Raidl, editors, *Applications of Evolutionary Computing, Proceedings of EvoWorkshops2002: EvoCOP, EvoIASP, EvoSTim*, volume 2279 of *LNCS*, pages 121–130. Springer, 2002.

M. Pelillo. *Encyclopedia of Optimization*, chapter Heuristics for Maximum Clique and Independent Set, pages 1508–1520. Springer, 2nd edition, 2009.

C. Perea, J. Alcaca, V. Yepes, F. Gonzalez-Vidosa, and A. Hospitaler. Design of reinforced concrete bridge frames by heuristic optimization. *Advances in Engineering Software*, 39(8):676–688, 2008.

S. Petrovic and Y. Bykov. A multiobjective optimisation approach for exam timetabling based on trajectories. In E. Burke and P. De Causmaecker, editors, *Practice and Theory of Automated Timetabling (PATAT) IV*, volume 2740 of *LNCS*, pages 181–194. Springer, 2003.

D. Porumbel, J.-K. Hao, and P. Kuntz. An evolutionary approach with diversity guarantee and well-informed grouping recombination for graph coloring. *Computers and Operations Research*, 37:1822–1832, 2010.

B. Reed. A strengthening of Brooks' theorem. *Journal of Combinatorial Theory, Series B*, 76(2):136–149, 1999.

C. Ribeiro and S. Urrutia. Heuristics for the mirrored travelling tournament problem. *European Journal of Operational Research*, 179(3):775–787, 2007.

A. Rizzoli, R. Montemanni, E. Lucibello, and L. Gambardella. Ant colony optimization for real-world vehicle routing problems. *Swarm Intelligence*, 1(2):135–151, 2007.

N. Robertson, D. Sanders, P. Seymour, and R. Thomas. The four color theorem. *Journal of Combinatorial Theory, Series B*, 70:2–44, 1997.

D. Rose, G. Lueker, and R. Tarjan. Algorithmic aspects of vertex elimination on graphs. *SIAM Journal of Computing*, 5(2):266–283, 1976.

P. Ross, E. Hart, and D. Corne. Genetic algorithms and timetabling. In A. Ghosh and K. Tsutsui, editors, *Advances in Evolutionary Computing: Theory and Applications*, Natural Computing, pages 755–771. Springer, 2003.

O. Rossi-Doria, M. Samples, M. Birattari, M. Chiarandini, J. Knowles, M. Manfrin, M. Mastrolilli, L. Paquete, B. Paechter, and T. Stützle. A comparison of the performance of different metaheuristics on the timetabling problem. In E. Burke and P. De Causmaecker, editors, *Practice and Theory of Automated Timetabling (PATAT) IV*, volume 2740 of *LNCS*, pages 329–351. Springer, 2002.

E. Russell and F. Jarvis. There are 5,472,730,538 essentially different Sudoku grids. Online Article: www.afjarvis.staff.shef.ac.uk/sudoku/sudgroup.html, September 2005.

K. Russell. Balancing carry-over effects in round-robin tournaments. *Biometrika*, 67(1):127–131, 1980.

R. Russell and J. Leung. Devising a cost effective schedule for a baseball league. *Operations Research*, 42(4):614–625, 1994.

A. Schaerf. A survey of automated timetabling. *Artificial Intelligence Review*, 13 (2):87–127, 1999.

S. Sekiner and M. Kurt. A simulated annealing approach to the solution of job rotation scheduling problems. *Applied Mathematics and Computation*, 188(1): 31–45, 2007.

K. Smith-Miles, D. Baatar, B. Wreford, and R. Lewis. Towards objective measures of algorithm performance across instance space. *Computers and Operations Research*, 45:12–24, 2014.

J. Spinrad and G. Vijayan. Worst case analysis of a graph coloring algorithm. *Discrete Applied Mathematics*, 12:89–92, 1984.

J. Thompson and K. Dowsland. An improved ant colony optimisation heuristic for graph colouring. *Discrete Applied Mathematics*, 156:313–324, 2008.

M. Trick. A schedule-then-break approach to sports timetables. In E. Burke and W. Erben, editors, *Practice and Theory of Automated Timetabling (PATAT) III*, volume 2079 of *LNCS*, pages 242–253. Springer, 2001.

J. Turner. Almost all k-colorable graphs are easy to color. *Journal of Algorithms*, 9: 63–82, 1988.

C. Valenzuela. A study of permutation operators for minimum span frequency assignment using an order based representation. *Journal of Heuristics*, 7:5–21, 2001.

J. van den Broek and C. Hurkens. An IP-based heuristic for the post enrolment course timetabling problrm of the ITC2007. *Annals of Operational Research*, 194:439–454, 2012.

P. van Laarhoven and E. Aarts. *Simulated Annealing: Theory and Applications*. Kluwer Academic Publishers, 1987.

V. Vizing. On an estimate of the chromatic class of a *p*-graph. *Diskret. Analiz.*, 3: 25–30, 1964.

R. Wilson. *Four Colors Suffice: How the Map Problem Was Solved*. Penguin Books, 2003.

L. Wolsey. *Integer Programming*. Wiley-Interscience, 1998.

M. Wright. Timetabling county cricket fixtures using a form of tabu search. *Journal of the Operational Research Society*, 47(7):758–770, 1994.

M. Wright. Scheduling fixtures for Basketball New Zealand. *Computers and Operations Research*, 33(7):1875–1893, 2006.

Q. Wu and J-K. Hao. Coloring large graphs based on independent set extraction. *Computers and Operational Research*, 39:283–290, 2012.

T. Yato and T. Seta. Complexity and completeness of finding another solution and its application to puzzles. *IEICE Transactions on Fundamentals of Electronics, Communications and Computer Sciences*, E86-A:1052–1060, 2003.

# Index

© Springer International Publishing Switzerland 2016
R.M.R. Lewis, *A Guide to Graph Colouring*,
DOI 10.1007/978-3-319-25730-3

Printed in the United States
By Bookmasters